AF774276

Lecture Notes in Physics

Founding Editors: W. Beiglböck, J. Ehlers, K. Hepp, H. Weidenmüller

Editorial Board

R. Beig, Vienna, Austria
W. Beiglböck, Heidelberg, Germany
W. Domcke, Garching, Germany
B.-G. Englert, Singapore
U. Frisch, Nice, France
F. Guinea, Madrid, Spain
P. Hänggi, Augsburg, Germany
W. Hillebrandt, Garching, Germany
R. L. Jaffe, Cambridge, MA, USA
W. Janke, Leipzig, Germany
H. v. Löhneysen, Karlsruhe, Germany
M. Mangano, Geneva, Switzerland
J.-M. Raimond, Paris, France
M. Salmhofer, Heidelberg, Germany
D. Sornette, Zurich, Switzerland
S. Theisen, Potsdam, Germany
D. Vollhardt, Augsburg, Germany
W. Weise, Garching, Germany
J. Zittartz, Köln, Germany

The Lecture Notes in Physics

The series Lecture Notes in Physics (LNP), founded in 1969, reports new developments in physics research and teaching – quickly and informally, but with a high quality and the explicit aim to summarize and communicate current knowledge in an accessible way. Books published in this series are conceived as bridging material between advanced graduate textbooks and the forefront of research and to serve three purposes:

- to be a compact and modern up-to-date source of reference on a well-defined topic
- to serve as an accessible introduction to the field to postgraduate students and nonspecialist researchers from related areas
- to be a source of advanced teaching material for specialized seminars, courses and schools

Both monographs and multi-author volumes will be considered for publication. Edited volumes should, however, consist of a very limited number of contributions only. Proceedings will not be considered for LNP.

Volumes published in LNP are disseminated both in print and in electronic formats, the electronic archive being available at springerlink.com. The series content is indexed, abstracted and referenced by many abstracting and information services, bibliographic networks, subscription agencies, library networks, and consortia.

Proposals should be sent to a member of the Editorial Board, or directly to the managing editor at Springer:

Christian Caron
Springer Heidelberg
Physics Editorial Department I
Tiergartenstrasse 17
69121 Heidelberg / Germany
christian.caron@springer.com

C. Kollmitzer
M. Pivk (Eds.)

Applied Quantum Cryptography

Springer

Christian Kollmitzer
AIT Austrian Institute
of Technology GmbH
Safety & Security Department
Quantum Technologies
Lakeside B01A, 9020, Klagenfurt
Austria
christian.kollmitzer@ait.ac.at

Mario Pivk
Pöckau 171
9601 Arnoldstein
Austria
mpivk@edu.uni-klu.ac.at

Kollmitzer C., Pivk M. (Eds.), *Applied Quantum Cryptography*, Lect. Notes Phys. 797 (Springer, Berlin Heidelberg 2010), DOI 10.1007/978-3-642-04831-9

Lecture Notes in Physics ISSN 0075-8450 e-ISSN 1616-6361
ISBN 978-3-642-04829-6 e-ISBN 978-3-642-04831-9
DOI 10.1007/978-3-642-04831-9
Springer Heidelberg Dordrecht London New York

Library of Congress Control Number: 2010920541

© Springer-Verlag Berlin Heidelberg 2010
This work is subject to copyright. All rights are reserved, whether the whole or part of the material is concerned, specifically the rights of translation, reprinting, reuse of illustrations, recitation, broadcasting, reproduction on microfilm or in any other way, and storage in data banks. Duplication of this publication or parts thereof is permitted only under the provisions of the German Copyright Law of September 9, 1965, in its current version, and permission for use must always be obtained from Springer. Violations are liable to prosecution under the German Copyright Law.
The use of general descriptive names, registered names, trademarks, etc. in this publication does not imply, even in the absence of a specific statement, that such names are exempt from the relevant protective laws and regulations and therefore free for general use.

Cover design: Integra Software Services Pvt. Ltd., Pondicherry

Printed on acid-free paper

Springer is part of Springer Science+Business Media (www.springer.com)

CK: For my family and Verena
MP: For all those, who enrich my life

Foreword

Using the quantum properties of single photons to exchange binary keys between two partners for subsequent encryption of secret data is an absolutely novel technology. Only a few years ago quantum cryptography – or better Quantum Key Distribution – was the domain of basic research laboratories at universities. But during the last few years things changed. Quantum Key Distribution or QKD left the laboratories and was picked up by more practical-oriented teams that worked hard to develop a practically applicable technology out of the astonishing results of basic research.

One major milestone toward a QKD technology was a large research and development project funded by the European Commission that aimed at combining quantum physics with complementary technologies that are necessary to create a technical solution: electronics, software, and network components were added within the project SECOQC (Development of a Global Network for Secure Communication based on Quantum Cryptography) that teamed up all expertise on European level to get a technology for future cryptography.

Lead-managed by a team at the Austrian Research Centers in Vienna, the practical application of QKD in a standard optical fiber network was demonstrated giving a glimpse of the future of secure communication. Although many steps have still to be done in order to achieve a real mature technology the cornerstone for future secure communication is already laid. QKD will not be the Holy Grail of security, it will not be able to solve all problems for evermore. But QKD has the potential to replace one of the weakest parts of symmetric encryption: the exchange of the key. It can be proven that the key exchange process cannot be corrupted and that keys that are generated and exchanged quantum cryptographically will be secure for ever (as long as some additional conditions are kept).

This book will show the state of the art of Quantum Cryptography and it will sketch how it can be implemented in standard communication infrastructure. The growing vulnerability of sensitive data requires new concepts and QKD will be a possible solution to overcome some of today's limitations.

Vienna, Austria Christian Monyk

Acknowledgements

We would like to give thanks to the Austrian Research Centers GmbH – ARC Klagenfurt and Vienna for their support of this book. This work was supported by the EC/IST Integrated Project SECOQC (contract no. 506813). M.S. is grateful to T. Länger, T. Lorünser, C. Pacher, M. Peev, and A. Poppe for discussion, help, and assistance in writing Chap. 6. We would like to take this opportunity to express our gratitude to Roland Potzmann from the Central Institute for Meteorology and Geodynamics (ZAMG) who provided us with the climate data. Furthermore, a special thanks to the people of the University of Vienna for their support. Christian Kollmitzer would like to give thanks to Mr. Gerald Dissauer for discussions and explanations of Medical Information Systems. He would also like to give thanks to Michele Mosca, Norbert Lütkenhaus, and Daniel Gottesman from the University of Waterloo, Ontario, Canada, and to Takashi Linzbichler from the University of Applied Science Joanneum, Kapfenberg, Austria, and their students for many hours of discussion on the "Ring of Trust" model. We want to thank Claus Ascheron for his work in the process of making this book and our reviewers for their fruitful comments.

Contents

Chapter 1
Introduction

C. Kollmitzer

Quantum cryptography or more precisely quantum key distribution (QKD) is a new technology which gets a high level of attention today worldwide. The possibility to exchange information in a provable secure way is a milestone in communication history. The main problem in QKD is the range limitation between the communication partners Alice and Bob. Several experiments have shown that the distance between Alice and Bob can be enhanced. Some of these experiments used optical fibers; others were based on free space technologies. But beyond that it is now possible to build communication networks based on QKD. Thus, not only Peer to Peer connection has to be used but it is now possible to build modern communication structures.

The first fully functional QKD-based network was presented in October 2008 in Vienna, Austria. It acted thereby as the base layer for a video conferencing network, which connected several parts of the city. Not only one QKD technology was used but five different systems were deployed. Single communications used several of them, invisible for the user.

This book contains the following aspects:

At first fundamental techniques are discussed which form a basis for all further concepts, in particular the QKD systems also introduced here. The individual steps of the communication setup are shown in detail, especially sifting, reconciliation, error correction, and privacy amplification.

Regarding error correction the Adaptive Cascade protocol, an improvement of the original error correction protocol Cascade, is introduced. It enhances the classical Cascade approach and supplements it with a method to determine the optimal initial block size and hence enhance its efficiency.

In order to ensure the security of communication systems, different attack strategies must be examined. Besides the classical attack strategies QKD systems offer a range of new aspects which are also presented.

C. Kollmitzer (✉)
Safety & Security Department, Quantum Technologies, AIT Austrian Institute of Technology GmbH, Lakeside B01A 9020 Klagenfurt, Austria,
`christian.kollmitzer@ait.ac.at; http://www.ait.ac.at`

Kollmitzer, C.: *Introduction*. Lect. Notes Phys. **797**, 1–2 (2010)
DOI 10.1007/978-3-642-04831-9_1 © Springer-Verlag Berlin Heidelberg 2010

After that we present current QKD systems in detail, which were used in the SECOQC network project of the European Union and were part of the first QKD-based network in October 2008 in Vienna, Austria.

Although QKD system has been used for years within different experimental setups, many of these experiments took place within a laboratory. But due to the deployment of the SECOQC network, it was possible to collect data from a longtime setup within an urban environment. The influence of the environment was examined and temperature, humidity, etc., could be seized for the first time and subsequently statistically evaluated. The results are discussed in detail.

QKD systems are designed as an enhancement to existing communication networks. Thus, their integration into current communication systems is a crucial factor. Special network protocols had been developed whereby in particular the Q3P protocol is of high importance and presented in detail.

A communication network is one of the key developments on the way to the field use. The appropriate fundamentals are presented. Apart from that we deal with the user itself and its benefit in using QKD networks. In particular the employments of QKD generated keys using current communication equipments like the iPhone are described.

Because of the range limitation of QKD systems, the development of global networks is one of the most strongly examined research areas. A possible solution based on a network of trusted communication centers is also presented. The main advantage in this model is the possibility to generate keys on demand such that the user doesn't need to store them in a relatively uncertain environment.

We hope that we can promote the interest in QKD systems and the associated new possibilities with this book. We assume today that these new options will be the subject of international research activities all over the next years worldwide and that the results will have a massive impact on the communication structures of tomorrow.

Chapter 2
Preliminaries

M. Pivk

This chapter discusses basics necessary for the next chapters. All fields are skimped, because some areas would need more explanation, like quantum information theory, but this will be out of scope for this chapter.

2.1 Quantum Information Theory

In this section a short introduction to quantum information is given. For detailed explanation we refer to the book of Nielsen and Chuang [4], where this topic is amplified.

2.1.1 Quantum Bits

Since Shannon and the beginning of information theory, the *bit* has been the basic term in classical information. The states of a bit are either 0 or 1. In accordance with the classical concept in quantum information exists the *qubit* (short for quantum bit). Like for the classical bit two states are possible, $|0\rangle$ and $|1\rangle$. This special notation '$|\rangle$' is called the *Dirac notation* (or *ket*) and is the standard notation for states in quantum mechanics. The major difference to the classical bit, which accepts only 0 or 1, is that a qubit also allows states in between $|0\rangle$ and $|1\rangle$, which are called *superpositions*. Let us denote this by

$$|\psi\rangle = \alpha|0\rangle + \beta|1\rangle, \tag{2.1}$$

where $\alpha, \beta \in \mathbb{C}$. Because these factors are complex numbers the state of a qubit can be described as a vector in a two-dimensional complex vector space $\mathbb{C}^2$, also called *Hilbert space*. The states $|0\rangle$ and $|1\rangle$ form the computational basis and are

M. Pivk (✉)
Safety & Security Department, Quantum Technologies, AIT Austrian Institute of Technology GmbH, Lakeside B01A, 9020 Klagenfurt, Austria, mario.pivk@ait.ac.at; http://www.ait.ac.at

Pivk, M.: *Preliminaries*. Lect. Notes Phys. **797**, 3–21 (2010)
DOI 10.1007/978-3-642-04831-9_2 © Springer-Verlag Berlin Heidelberg 2010

orthonormal (see Definition 2.4) to each other, e.g., $|0\rangle = \begin{pmatrix} 1 \\ 0 \end{pmatrix}$ and $|1\rangle = \begin{pmatrix} 0 \\ 1 \end{pmatrix}$. Since a qubit state is a unit vector, meaning the length is normalized to 1, following equation must be fulfilled by the scalars α, β:

$$|\alpha|^2 + |\beta|^2 = 1. \tag{2.2}$$

Using this fact we can rewrite the state of a qubit

$$|\psi\rangle = \cos\frac{\theta}{2}|0\rangle + e^{i\varphi}\sin\frac{\theta}{2}|1\rangle, \tag{2.3}$$

where θ, φ are real numbers and define a point on a sphere called the *Bloch sphere* (see Fig. 2.1).

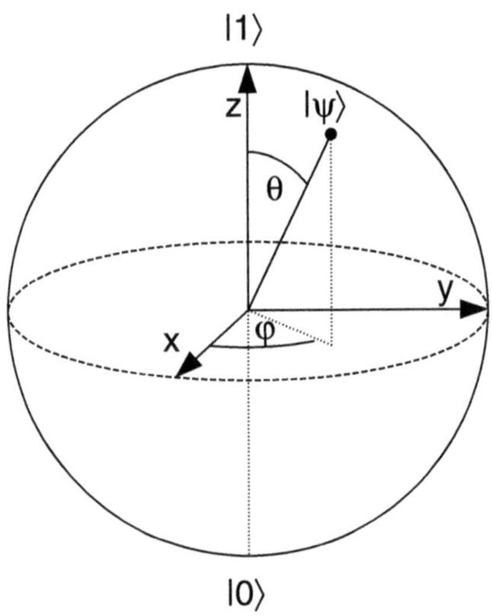

Fig. 2.1 Bloch sphere representation of a qubit

The measurement of qubits is a problem. In the special case when α or β is 0, the mapping to the classical bit will result in 1 or 0, respectively, as expected. But what happens if the qubit is in another superposition, i.e., $\alpha, \beta \neq 0$? Depending on the scalars the qubit will be measured as 1 with a certain probability or as 0 with the complementary probability. Since the scalars fulfil Eq. 2.2, the probability for a qubit to be measured as 0 is $|\alpha|^2$ and as 1 it is $|\beta|^2$. We see this in detail in Sect. 2.1.3.

Furthermore in quantum mechanics the scalars α and β are also called the *amplitudes* of the states $|0\rangle$ and $|1\rangle$, respectively. But there exists a second term describing a qubit, the *phase*. Consider the state $e^{i\varphi}|\psi\rangle$, where $|\psi\rangle$ is a state vector, and φ is a real number. We say that the state $e^{i\varphi}|\psi\rangle$ is equal to $|\psi\rangle$, up to the *global phase factor* $e^{i\varphi}$. The measurements for these two states are from the point of statistics the same as you will see in Sect. 2.1.2.

Another kind of phase is the *relative phase*. Consider these two states

$$|+\rangle = \frac{1}{\sqrt{2}}(|0\rangle + |1\rangle) \qquad \text{and} \qquad |-\rangle = \frac{1}{\sqrt{2}}(|0\rangle - |1\rangle). \tag{2.4}$$

In the state $|+\rangle$ the amplitude of $|1\rangle$ is $\frac{1}{\sqrt{2}}$. In state $|-\rangle$ the amplitude has the same magnitude but a different sign. We define that two amplitudes α_1, α_2 for some states differ by a *relative phase* if there is a real φ such that the $\alpha_1 = e^{i\varphi}\alpha_2$. In contrast to the *global phase*, where both amplitudes of the state are different by the factor $e^{i\varphi}$, the *relative phase* differs only in one amplitude by the factor $e^{i\varphi}$.

2.1.2 Linear Operators

The state change of qubits is done by linear operators. Therefore a function A is used, taking vectors from $\mathcal{V}$ to $\mathcal{W}$ ($\mathcal{V}$ and $\mathcal{W}$ are vector spaces of $\mathbb{C}^*$). The most convenient way to describe such a function is the *matrix representation*. If matrix A has m columns and n rows and this matrix is multiplied with the vector $|v\rangle \in \mathbb{C}^n$ we get a new vector $|w\rangle \in \mathbb{C}^m$ as result. The claim for such a matrix A is to fulfill the linearity equation [4]

$$A\left(\sum_i a_i |v_i\rangle\right) = \sum_i a_i A|v_i\rangle. \tag{2.5}$$

Let $A : \mathcal{V} \longrightarrow \mathcal{W}$ be a linear operator and $|v_1\rangle, \ldots, |v_n\rangle$ be a basis of $\mathcal{V}$ and $|w_1\rangle, \ldots, |w_m\rangle$ a basis of $\mathcal{W}$. There exist complex numbers $A_{1j}, \ldots, A_{mj}$,

$$A|v_j\rangle = \sum_i A_{ij}|w_i\rangle \qquad \text{with} \qquad 1 \le i \le m, 1 \le j \le n, \tag{2.6}$$

which form the matrix representation of the operator A.

Contrarily, a $n \times m$ matrix can be understood as the opposite linear operator sending vectors out of the vector space $\mathcal{W}$ to the vector space $\mathcal{V}$ by performing the matrix multiplication with those vectors.

We use a notation which is different to the usual notation in linear algebra. Table 2.1 lists some frequently used symbols in quantum mechanics. As we know, a vector can be represented by the sum of the vectors out of the computational basis. For simplification we take the computational basis $v_1 = \begin{pmatrix} 1 \\ 0 \\ \vdots \\ 0 \end{pmatrix}$, $v_2 = \begin{pmatrix} 0 \\ 1 \\ \vdots \\ 0 \end{pmatrix}, \ldots,$

Table 2.1 Summary of some standard quantum mechanical notation

Notation	Description
z^*	Complex conjugate of the complex number z. e.g., $(1+i)^* = 1-i$
$\|v\rangle$	Vector. Also known as a *ket*. $\|v\rangle = \sum_i a_i \|v_i\rangle$
$\langle v\|$	Vector dual to $\|v\rangle$. Also known as a *bra*. $\langle v\| = \sum_i a_i^* \|v_i\rangle^T$
$\lambda\|v\rangle$	Multiplication by a scalar λ. $\lambda\|v\rangle = \sum_i \lambda a_i \|v_i\rangle$
$\langle v\|w\rangle$	Inner product between the vectors $\|v\rangle$ and $\|w\rangle$
$\|v\rangle\langle w\|$	Outer product of $\|v\rangle$ and $\|w\rangle$
$\|v\rangle \otimes \|w\rangle$	Tensor product of $\|v\rangle$ and $\|w\rangle$
A^*	Complex conjugate of the A matrix
A^T	Transpose of the A matrix
$A^\dagger$	Hermitian conjugate or ad-joint of the A matrix, $A^\dagger = (A^T)^*$
$\langle\varphi\|A\|\psi\rangle$	Inner product between $\|\varphi\rangle$ and $A\|\psi\rangle$

$$v_n = \begin{pmatrix} 0 \\ 0 \\ \vdots \\ 1 \end{pmatrix} \text{ such that a vector } |v\rangle = \textstyle\sum_i a_i |v_i\rangle \text{ can also be written as } |v\rangle = \begin{pmatrix} a_1 \\ \vdots \\ a_n \end{pmatrix}.$$

If an other computational basis is used, it is written explicitly.

2.1.2.1 The Pauli Matrices

Four extremely useful matrices are the *Pauli matrices*. These are 2 by 2 matrices and represent some needed effects on qubits. The matrices are

$$X = \begin{pmatrix} 0 & 1 \\ 1 & 0 \end{pmatrix} \qquad Y = \begin{pmatrix} 0 & -i \\ i & 0 \end{pmatrix} \qquad Z = \begin{pmatrix} 1 & 0 \\ 0 & -1 \end{pmatrix}. \tag{2.7}$$

The fourth matrix is the identity matrix (I). The Pauli operators X and Z are also known as *bit flip* and *phase flip* operators. If we apply the X operation on a qubit we see that $|0\rangle$ changes to $|1\rangle$ and vice versa, i.e.,

$$X|0\rangle = \begin{pmatrix} 0 & 1 \\ 1 & 0 \end{pmatrix}\begin{pmatrix} 1 \\ 0 \end{pmatrix} = \begin{pmatrix} 0 \\ 1 \end{pmatrix} \quad \text{and} \quad X|1\rangle = \begin{pmatrix} 0 & 1 \\ 1 & 0 \end{pmatrix}\begin{pmatrix} 0 \\ 1 \end{pmatrix} = \begin{pmatrix} 1 \\ 0 \end{pmatrix}.$$

The Z operator is called phase flip operator because it changes the phase of $|1\rangle$ by the sign, i.e.,

$$Z|+\rangle = \begin{pmatrix} 1 & 0 \\ 0 & -1 \end{pmatrix}\begin{pmatrix} \sqrt{2} \\ \sqrt{2} \end{pmatrix} = \begin{pmatrix} \sqrt{2} \\ -\sqrt{2} \end{pmatrix} \quad \text{and}$$

$$Z|-\rangle = \begin{pmatrix} 1 & 0 \\ 0 & -1 \end{pmatrix}\begin{pmatrix} \sqrt{2} \\ -\sqrt{2} \end{pmatrix} = \begin{pmatrix} \sqrt{2} \\ \sqrt{2} \end{pmatrix}$$

A possibility to illustrate Y is to multiply the matrix with the imaginary unit i so we deal only with natural-numbered matrices. Thus, the reformulated version of Y is

$$iY = \begin{pmatrix} 0 & 1 \\ -1 & 0 \end{pmatrix}.$$

The iY operator performs both flips, a bit flip and a phase flip, since

$$iY = ZX = \begin{pmatrix} 1 & 0 \\ 0 & -1 \end{pmatrix} \begin{pmatrix} 0 & 1 \\ 1 & 0 \end{pmatrix} = \begin{pmatrix} 0 & 1 \\ -1 & 0 \end{pmatrix} = i \begin{pmatrix} 0 & -i \\ i & 0 \end{pmatrix}.$$

Therefore, when the iY operator is applied on the states $|0\rangle$ and $|1\rangle$ we get

$$iY|0\rangle = \begin{pmatrix} 0 & 1 \\ -1 & 0 \end{pmatrix} \begin{pmatrix} 1 \\ 0 \end{pmatrix} = \begin{pmatrix} 0 \\ -1 \end{pmatrix} \quad \text{and} \quad iY|1\rangle = \begin{pmatrix} 0 & 1 \\ -1 & 0 \end{pmatrix} \begin{pmatrix} 0 \\ 1 \end{pmatrix} = \begin{pmatrix} 1 \\ 0 \end{pmatrix}.$$

2.1.2.2 Inner Products

An *inner product* (or *scalar product*) $\langle v|w\rangle$ (usually notation in linear algebra $(|v\rangle, |w\rangle)$) is a function which takes as input two vectors $|v\rangle$ and $|w\rangle$ from vector space $\mathcal{V}$ and produces a complex number as output. For example, the inner product of two n-dimensional vectors over the field of complex numbers is defined as

$$\langle v|w\rangle = \sum_i a_i^* b_i = \left(a_1^* \cdots a_n^*\right) \cdot \begin{pmatrix} b_1 \\ \vdots \\ b_n \end{pmatrix}. \tag{2.8}$$

The inner product satisfies the following requirements:

1. It is linear in the second argument

$$\left(|v\rangle, \sum_i \lambda_i |w_i\rangle\right) = \sum_i \lambda_i \left(|v\rangle, |w_i\rangle\right).$$

2. $\langle v|w\rangle = \langle w|v\rangle^*$.
3. $\langle v|v\rangle \geq 0$ with equality if and only if $|v\rangle = 0$.

In the following some definitions in connection with the inner product are given.

Definition 2.1 Let $\mathcal{V}$ be a set of vectors over $\mathbb{C}^n$. Then two vectors $|v\rangle, |w\rangle \in \mathcal{V}$ are *orthogonal*, if their inner product is 0.

Definition 2.2 Let $\mathcal{V}$ be a set of vectors over $\mathbb{C}^n$. The norm of a vector $|v\rangle \in \mathcal{V}$ is defined by $\||v\rangle\| = \sqrt{\langle v|v\rangle}$. The norm of a vector is often understood as its length or size.

Definition 2.3 Let $\mathcal{V}$ be a set of vectors over $\mathbb{C}^n$. Then a vector $|v\rangle \in \mathbb{V}$ is called a unit vector or the vector is normalized if $\||v\rangle\| = 1$. Normalizing a vector means dividing it by its norm:

$$\left\| \frac{|v\rangle}{\||v\rangle\|} \right\| = 1.$$

Definition 2.4 Let $\mathcal{V}$ be a set of vectors over $\mathbb{C}^n$. Then a subset of vectors $|v_i\rangle \in \mathbb{V}$ is called orthonormal if each vector $|v_i\rangle$ is a unit vector, and distinct vectors are orthogonal $\langle v_i|w_j\rangle = 0,\ i, j = 1...n,\ i \neq j$.

For the computational basis of a vector space the last definition of orthonormal must hold. So those vectors form the spanning set for the vector space and any vector out of this space can be written as a linear combination.

2.1.2.3 Outer Products

The *outer product* of two vectors is the contrary multiplication to the inner product. In opposite to the inner product resulting in a single complex value, the outer product yields to a matrix:

$$|v\rangle\langle w| = A_{i,j} = \begin{pmatrix} a_1 \\ \vdots \\ a_m \end{pmatrix} \cdot \begin{pmatrix} b_1^* \cdots b_n^* \end{pmatrix} = \begin{pmatrix} a_1 b_1^* & \cdots & a_1 b_n^* \\ \vdots & \ddots & \vdots \\ a_m b_1^* & \cdots & a_m b_n^* \end{pmatrix}. \tag{2.9}$$

The outer product representation is a useful way of representing linear operators which makes use of the inner product. Let $|v\rangle$ be a vector in an inner product space $\mathcal{V}$ and $|w\rangle$ be a vector in an inner product space $\mathcal{W}$. Define $|w\rangle\langle v|$ to be the linear operator from $\mathcal{V}$ to $\mathcal{W}$ like

$$(|w\rangle\langle v|)(|v'\rangle) = |w\rangle\langle v|v'\rangle = \langle v|v'\rangle|w\rangle. \tag{2.10}$$

The equation imposes at least two interpretations. On the one hand the vector $|v'\rangle$ is mapped by the matrix to a vector which lies in $\mathcal{W}$; on the other hand, it is only the representation of the vector $|w\rangle$ multiplied by a complex value.

One application of the outer product notation can be discerned from an important result known as the completeness relation for orthonormal vectors. Let $|v_i\rangle$ be orthonormal basis for the vector space $\mathcal{V}$. Then following equation must be fulfilled

$$\sum_i |v_i\rangle\langle v_i| = I. \tag{2.11}$$

2.1.2.4 Tensor Products

The *tensor product* is an operation to create a larger vector space from two smaller vector spaces. We have two vector spaces V and W of dimensions m and n, respectively. Then $V \otimes W$ is an mn dimensional vector space, whose elements are linear combinations of tensor products of elements $|v\rangle \in V$ and $|w\rangle \in W$.

For example, the tensor product of vectors $(1, 2)$ and $(3, 4)$ is the vector

$$\begin{pmatrix} 1 \\ 2 \end{pmatrix} \otimes \begin{pmatrix} 3 \\ 4 \end{pmatrix} = \begin{pmatrix} 1 \times 3 \\ 1 \times 4 \\ 2 \times 3 \\ 2 \times 4 \end{pmatrix} = \begin{pmatrix} 3 \\ 4 \\ 6 \\ 8 \end{pmatrix}.$$

In the previous example, we perform the operation on two vector spaces, but the tensor product can also be applied on the linear operators of vector spaces. Assume $A : V \to V'$ and $B : W \to W'$ then $A \otimes B : V \otimes W \to V' \otimes W'$. Suppose A is a m by n matrix, and B is a p by q matrix. Then we have the matrix representation:

$$A \otimes B \equiv \underbrace{\left.\begin{pmatrix} A_{11}B & A_{12}B & \cdots & A_{1n}B \\ A_{21}B & A_{22}B & \cdots & A_{2n}B \\ \vdots & \vdots & \vdots & \vdots \\ A_{m1}B & A_{m2}B & \cdots & A_{mn}B \end{pmatrix}\right\}}_{nq} mp, \tag{2.12}$$

where $A_{11}B$ denotes a p by q submatrix. For example, the tensor product of the Pauli matrices X and Y is

$$X \otimes Y = \begin{pmatrix} 0 \cdot Y & 1 \cdot Y \\ 1 \cdot Y & 0 \cdot Y \end{pmatrix} = \begin{pmatrix} 0 & 0 & 0 & -i \\ 0 & 0 & i & 0 \\ 0 & -i & 0 & 0 \\ i & 0 & 0 & 0 \end{pmatrix}.$$

2.1.3 Quantum Measurement

This section provides ways for describing the effects of measurements on quantum systems with reference to the four postulates in [4]. Before we can measure a quantum we have to set up the area in which quantum mechanics takes place.

Postulate 1 of [4]: A complex vector space with inner product (also called Hilbert space) is related with any isolated physical system. This is also known as the state space of the system. With the unit vectors of the system's state space (state vectors) we can span the complete system.

To get more information of a particular system we would measure the state space. But not in quantum mechanics, here we cannot measure what the state space of the system is, nor we can tell what the state vector of that system is. The simplest and most important system is the *qubit*, described in Sect. 2.1.1. The next postulate gives the description how states change with time.

Postulate 2 of [4]: The evolution of a *closed* quantum system is described by a *unitary transformation*. That is, the state $|\psi\rangle$ of the system at time t_1 is related to the state $|\psi'\rangle$ of the system at time t_2 by a unitary operator U which depends only on the times t_1 and t_2,

$$|\psi'\rangle = U|\psi\rangle. \tag{2.13}$$

After quantum mechanics does not tell us the state space and quantum state of a system, it only assures us which unitary operators U describe the change in any closed quantum system. Such operators we have already seen in Sect. 2.1.2. Setting up the base, we can continue with the measurement.

Postulate 3 of [4]: Quantum measurements are described by a collection $\{M_m\}$ of *measurement operators*. These are operators acting on the state space of the system being measured. The index m refers to the measurement outcomes that may occur in the experiment. If the state of the quantum system is $|\psi\rangle$ immediately before the measurement then the probability that result m occurs is given by

$$p(m) = \langle\psi|M_m^\dagger M_m|\psi\rangle, \tag{2.14}$$

and the state of the system after the measurement is

$$\frac{M_m|\psi\rangle}{\sqrt{\langle\psi|M_m^\dagger M_m|\psi\rangle}}. \tag{2.15}$$

The measurement operators satisfy the completeness equation

$$\sum_m M_m^\dagger M_m = I. \tag{2.16}$$

The completeness equation expresses the fact that the probabilities sum to 1:

$$\sum_m p(m) = \sum_m \langle\psi|M_m^\dagger M_m|\psi\rangle = 1. \tag{2.17}$$

There are different types of measurements, but the most important in our case is the measurement in the computational basis. Hence, we know that $|0\rangle$ and $|1\rangle$ form a computational basis for the two-dimensional complex vector space (space of qubits). As said in a previous section we can map these states onto the states 0 and 1 of a classical bit during measurement. Now we define two measurement operators M_0, M_1:

$$M_0 = |0\rangle\langle 0| = \begin{pmatrix} 1 & 0 \\ 0 & 0 \end{pmatrix} \quad \text{and} \quad M_1 = |1\rangle\langle 1| = \begin{pmatrix} 0 & 0 \\ 0 & 1 \end{pmatrix}. \tag{2.18}$$

Observe that for the operators apply $M_0^\dagger = M_0$, $M_1^\dagger = M_1$ and $M_0^2 = M_0$, $M_1^2 = M_1$. A measurement on a qubit with state $|\psi\rangle = \alpha|0\rangle + \beta|1\rangle$. If we use the operator M_0 for the qubit we obtain a probability that the result is 0

$$\begin{aligned} p(0) &= \langle\psi|M_0^\dagger M_0|\psi\rangle \\ &= \langle\psi|M_0|\psi\rangle \\ &= \begin{pmatrix}\alpha^* & \beta^*\end{pmatrix}\begin{pmatrix}1 & 0\\ 0 & 0\end{pmatrix}\begin{pmatrix}\alpha\\ \beta\end{pmatrix} \\ &= \begin{pmatrix}\alpha^* & \beta^*\end{pmatrix}\begin{pmatrix}\alpha\\ 0\end{pmatrix} \\ &= \alpha^*\alpha + 0 = |\alpha|^2. \end{aligned} \tag{2.19}$$

Similarly, we get the probability $p(1) = |\beta|^2$ for the measurement result 1. The state of the system after the measurement is

$$\begin{aligned} |\psi'\rangle &= \frac{M_0|\psi\rangle}{\sqrt{\langle\psi|M_0^\dagger M_0|\psi\rangle}} = \frac{\begin{pmatrix}1 & 0\\ 0 & 0\end{pmatrix}\begin{pmatrix}\alpha\\ \beta\end{pmatrix}}{\sqrt{|\alpha|^2}} \\ &= \frac{\begin{pmatrix}\alpha\\ 0\end{pmatrix}}{|\alpha|} = \frac{\alpha|0\rangle}{|\alpha|} = \frac{\alpha}{|\alpha|}|0\rangle \end{aligned} \tag{2.20}$$

if the result was 0. Analog the state

$$|\psi'\rangle = \frac{\beta}{|\beta|}|1\rangle \tag{2.21}$$

for the measurement result 1.

Based on Eq. 2.4 if the qubit is in the specific state $|+\rangle$ we have $\alpha = \beta = \frac{1}{\sqrt{2}}$ and using Eq. 2.14 we get 0 as well as 1 with probability

$$p(0) = p(1) = |\alpha|^2 = |\beta|^2 = \left|\frac{1}{\sqrt{2}}\right|^2 = \frac{1}{2}. \tag{2.22}$$

Due to Eq. 2.15 after the measurement the system's state is

$$\frac{\alpha}{|\alpha|}|0\rangle = \frac{\frac{1}{\sqrt{2}}}{\left|\frac{1}{\sqrt{2}}\right|}|0\rangle = |0\rangle \quad \text{or} \quad \frac{\beta}{|\beta|}|1\rangle = \frac{\frac{1}{\sqrt{2}}}{\left|\frac{1}{\sqrt{2}}\right|}|1\rangle = |1\rangle, \tag{2.23}$$

respectively. We get the same results for the state $|-\rangle$.

To perform a correct measurement for such states, we cannot use the computational basis $\{|0\rangle, |1\rangle\}$. Therefore we have to use the basis $\{|+\rangle, |-\rangle\}$ (defined in Eq. 2.4). We have seen that these states are orthonormal to each other, since $\langle +|+\rangle = \langle -|-\rangle = 1$ and $\langle +|-\rangle = 0$, so we can use them as computational basis.

As before, we map the two states $|+\rangle, |-\rangle$ onto classical bits, e.g., $|+\rangle \rightarrow 0$ and $|-\rangle \rightarrow 1$. Thus, the two operators are

$$M_0 = |+\rangle\langle+| = \frac{1}{2}\begin{pmatrix} 1 & 1 \\ 1 & 1 \end{pmatrix} \quad \text{and} \quad M_1 = |-\rangle\langle-| = \frac{1}{2}\begin{pmatrix} 1 & -1 \\ -1 & 1 \end{pmatrix}. \tag{2.24}$$

Using these measurement operators and the states $|+\rangle$ and $|-\rangle$ of Eq. 2.4 we get the probability for the result 0

$$p(0) = \langle +|M_0^\dagger M_0|+\rangle = \langle +|+\rangle\langle +|+\rangle = 1 \cdot 1 = 1. \tag{2.25}$$

So for the other case

$$p(1) = \langle +|M_1^\dagger M_1|+\rangle = \langle +|-\rangle\langle -|+\rangle = 0 \cdot 0 = 0, \tag{2.26}$$

and the state after the measurement is

$$\frac{M_0|+\rangle}{\sqrt{\langle +|M_0^\dagger M_0|+\rangle}} = \frac{|+\rangle\langle +|+\rangle}{\sqrt{\langle +|+\rangle\langle +|+\rangle}} = \frac{1}{1} \cdot |+\rangle = |+\rangle. \tag{2.27}$$

As the probability for the result 0 was 1 the state $|+\rangle$ is preserved even after the measurement.

With the computational basis $\{|0\rangle, |1\rangle\}$ and measurement states $|+\rangle, |-\rangle$, now the measurement of states $|0\rangle, |1\rangle$ in the computational basis $\{|+\rangle, |-\rangle\}$ yields the same probability results:

$$\begin{aligned} p(0) &= \langle 0|M_0^\dagger M_0|0\rangle = \langle 0|+\rangle\langle +|0\rangle = \frac{1}{\sqrt{2}} \cdot \frac{1}{\sqrt{2}} = \frac{1}{2}, \\ p(1) &= \langle 0|M_1^\dagger M_1|0\rangle = \langle 0|-\rangle\langle -|0\rangle = \frac{1}{\sqrt{2}} \cdot \frac{1}{\sqrt{2}} = \frac{1}{2}, \\ p(0) &= \langle 1|M_0^\dagger M_0|1\rangle = \langle 1|+\rangle\langle +|1\rangle = \frac{1}{\sqrt{2}} \cdot \frac{1}{\sqrt{2}} = \frac{1}{2}, \\ p(1) &= \langle 1|M_1^\dagger M_1|1\rangle = \langle 1|-\rangle\langle -|1\rangle = -\frac{1}{\sqrt{2}} \cdot -\frac{1}{\sqrt{2}} = \frac{1}{2}, \end{aligned} \tag{2.28}$$

and the state after the measurement is

$$\frac{M_0|0\rangle}{\sqrt{\langle 0|M_0^\dagger M_0|0\rangle}} = \frac{|+\rangle\langle+|0\rangle}{\sqrt{\langle 0|+\rangle\langle+|0\rangle}} = \frac{\frac{1}{\sqrt{2}}}{\frac{1}{\sqrt{2}}} \cdot |+\rangle = |+\rangle$$

or

$$\frac{M_1|0\rangle}{\sqrt{\langle 0|M_1^\dagger M_1|0\rangle}} = \frac{|-\rangle\langle-|0\rangle}{\sqrt{\langle 0|-\rangle\langle-|0\rangle}} = \frac{\frac{1}{\sqrt{2}}}{\frac{1}{\sqrt{2}}} \cdot |-\rangle = |-\rangle$$

and

$$\frac{M_0|1\rangle}{\sqrt{\langle 1|M_0^\dagger M_0|1\rangle}} = \frac{|+\rangle\langle+|1\rangle}{\sqrt{\langle 0|+\rangle\langle+|0\rangle}} = \frac{\frac{1}{\sqrt{2}}}{\frac{1}{\sqrt{2}}} \cdot |+\rangle = |+\rangle$$

or

$$\frac{M_1|1\rangle}{\sqrt{\langle 1|M_1^\dagger M_1|1\rangle}} = \frac{|-\rangle\langle-|1\rangle}{\sqrt{\langle 1|-\rangle\langle-|1\rangle}} = \frac{-\frac{1}{\sqrt{2}}}{\frac{1}{\sqrt{2}}} \cdot |-\rangle = (-1)\cdot|-\rangle = |-\rangle. \tag{2.29}$$

In Sect. 2.1.1 we said if qubits differ only by a global phase factor, they have the same statistical properties and so we consider them to be the same. Thus, in the last line in Eq. 2.29 we can neglect the factor -1.

2.1.4 The No-Cloning Theorem

Is it possible to make a copy of an unknown quantum state? The answer is no. In [10] the no-cloning theorem was presented the first time.

Suppose we have a quantum machine with two slots labeled A and B. Slot A is the *data slot* and starts out in a quantum state, $|\psi\rangle$. We do not know which state it has. The goal is to copy the state into slot B, the *target slot*. We assume that the target slot starts out in some independent state, $|s\rangle$. Thus the initial state of the copying machine is

$$|\psi\rangle \otimes |s\rangle. \tag{2.30}$$

Some unitary evolution U now effects the copying procedure, ideally,

$$|\psi\rangle \otimes |s\rangle \to_U U(|\psi\rangle \otimes |s\rangle) = |\psi\rangle \otimes |\psi\rangle. \tag{2.31}$$

Suppose this copying procedure works for two particular states, $|\psi\rangle$ and $|\varphi\rangle$. Then we have

$$U(|\psi\rangle \otimes |s\rangle) = |\psi\rangle \otimes |\psi\rangle, \tag{2.32}$$
$$U(|\varphi\rangle \otimes |s\rangle) = |\varphi\rangle \otimes |\varphi\rangle. \tag{2.33}$$

Taking the inner product of these two equations gives

$$\langle\psi|\varphi\rangle = (\langle\psi|\varphi\rangle)^2. \tag{2.34}$$

But $x = x^2$ has only two solution, $x = 0$ and $x = 1$, so either $|\psi\rangle = |\varphi\rangle$ or $|\psi\rangle$ and $|\varphi\rangle$ are orthogonal. If a cloning device would exist it can only clone states which are orthogonal to one another, and therefore a general quantum cloning device cannot exist. For example, we have qubits with states $|\psi\rangle$ and $|0\rangle$ where ψ is not $|1\rangle$, so it is impossible for a quantum cloner to copy them, since these states are not orthogonal.

2.2 Unconditional Secure Authentication

QKD communication is divided into two channels: the quantum channel and the classical public channel. On the public channel we have the problem that the adversary Eve can start a man-in-the-middle attack. So the need of authentication for the communication is essential. Usually authentication is done by public key methods, e.g., RSA [6] or DSS [3]. But this security is only computational. QKD's security is an unconditional secure, by using such an authentication scheme we will reduce it also to computational security. Therefore unconditional secure authentication schemes are necessary for QKD.

Next we present universal hashing, which can be used as an authentication scheme, where the authentication parties hold a pre-shared secret. This was first presented by Wegman and Carter [9]. Nowadays, more effective symmetric authentication methods are known. We will use Wegman–Carter authentication, because it describes an upper bound for needed symmetric authentication key.

2.2.1 *Universal Hashing*

The fundamental idea of universal hashing is to choose a hash function at random, independent of the input, such that the probability that any two distinct inputs have the same hash values is sufficiently low. To render more precisely, let $\mathcal{A}$ and $\mathcal{B}$ be finite sets, characterized by the number of elements $a = |\mathcal{A}|$ and $b = |\mathcal{B}|$, where $a \geq b$. The hash function h maps an element $x \in \mathcal{A}$ into an element $h(x) \in \mathcal{B}$. For h and for $x, y \in \mathcal{A}$, $x \neq y$, we define $\delta_h(x, y) = 1$ if $h(x) = h(y)$ and $\delta_h(x, y) = 0$ otherwise. Meaning $\delta_h(x, y) = 1$ if and only if the hashed values of x and y collide. The collection of all hash functions h, mapping the set $\mathcal{A}$ to set $\mathcal{B}$, is given as $\mathcal{H}$. Hence, the number of collisions for the hash function set $\mathcal{H}$ and the values x, y is defined as

$$\delta_{\mathcal{H}}(x, y) = \sum_{h \in \mathcal{H}} \delta_h(x, y).$$

Thus, the probability for the two distinct values can be computed by $\delta_{\mathcal{H}}(x, y)/|\mathcal{H}|$. The goal is to make this probability small, which can be achieved by a large number of hash functions $|\mathcal{H}|$. But when increasing $|\mathcal{H}|$ the number of bits for specifying the function $\log_2 |\mathcal{H}|$ increases as well. This case would be unpractical for applications since more random bits are needed. So we have to consider this when choosing $\mathcal{H}$.

Wegman and Carter [1, 9] were the first ones dealing with universal hashing and they defined some useful properties. Below, a definition of the term *universal* is given [8].

Definition 2.5 Let ε be a positive real number. $\mathcal{H}$ is ε-almost $universal_2$ (or ε-AU_2) if $\delta_{\mathcal{H}}(x, y) \leq \varepsilon|\mathcal{H}|$ for all $x, y \in \mathcal{A}, x \neq y$.

In other words the collision probability for any two inputs x and y is at most ε. Note that ε is bounded below by $1/b$. In the special case when $\varepsilon = 1/b$ we can skip the term *almost* and speak only about a $universal_2$ class of hash functions as Wegman and Carter [1] have done it. Next we define a stricter property for such classes [8].

Definition 2.6 Let ε be a positive real number. $\mathcal{H}$ is ε-almost strongly $universal_2$ (or ε-ASU_2) if

(a) for every $x_1 \in \mathcal{A}$ and for every $y_1 \in \mathcal{B}$, $|\{h \in \mathcal{H} : h(x_1) = y_1\}| = |\mathcal{H}|/|\mathcal{B}|$,
(b) for every $x_1, x_2 \in \mathcal{A}(x_1 \neq x_2)$ and for every $y_1, y_2 \in \mathcal{B}$,
$|\{h \in \mathcal{H} : h(x_1) = y_1, h(x_2) = y_2\}| \leq \varepsilon|\mathcal{H}|/|\mathcal{B}|$.

The first part of the definition says that any input x_1 has the probability $1/b$ to be mapped to a hashed value y_1 (see Fig. 2.2). The second part concerns the conditional probability that with given tuple (x_1, y_1) the probability that x_2 is mapped to y_2 is at most ε (see Fig. 2.3). As before in the special case we call it *strongly* $universal_2$ class of hash functions when $\varepsilon = 1/b$.

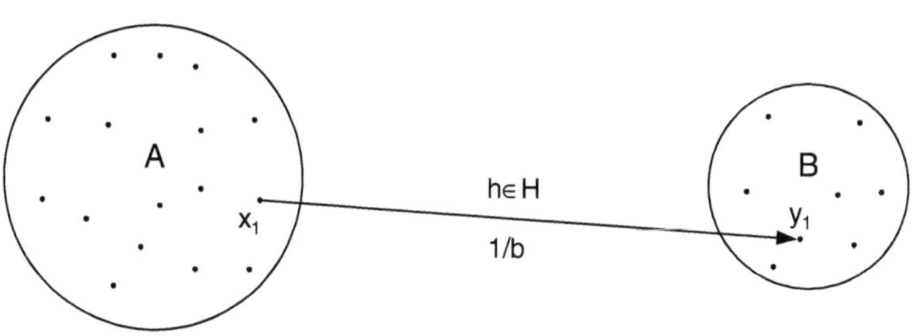

Fig. 2.2 Definition 2.6a. Any element of $\mathcal{A}$ is mapped to any element of $\mathcal{B}$ with probability $1/b$

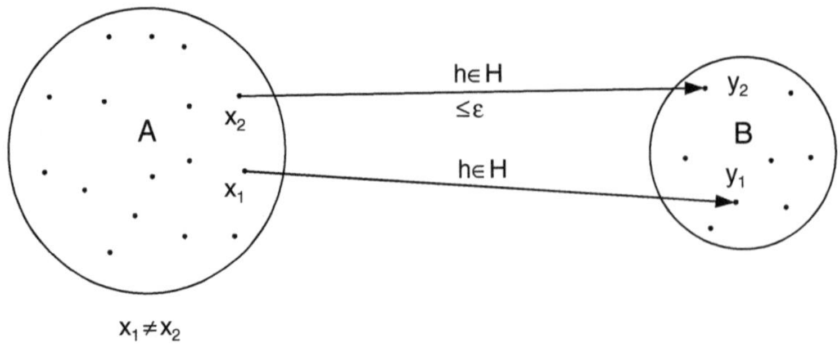

Fig. 2.3 Definition 2.6b. With given tuple (x_1, y_1) any x_2 is mapped to y_2 with probability $\leq \varepsilon$

2.2.2 Authentication

In the previous section we define what ε-*almost strongly universal*$_2$ families of hash function are. For authentication we can use them in the following manner: Alice and Bob share a secret random key k and have agreed on a ε-ASU set of hash function $\mathcal{H} = \{h|h : \mathcal{M} \rightarrow \mathcal{T}\}$ (not necessarily secret), where $\mathcal{M}$ is the set of possible messages, $\mathcal{T}$ the set of authentication tags, k has the length $\log_2(|\mathcal{H}|)$, and the hash function in $\mathcal{H}$ are ordered $\{h_0, h_1, ..., h_{|\mathcal{H}|-1}\}$. Alice sends Bob a message $m \in \mathcal{M}$ and the authentication tag $t = h_k(m)$. For Bob now it is easy to check if the message is really from Alice, he compares if the received tag t is equal to $h_k(m)$ and accepts the message as authentic if it does. The key k is then discarded. The authentication system's resistance against forgery has to be stated. A forger has two possibilities to attack the system: First, he can place a new message m' into the channel. Because the key k is random and secret, he does not know which hash function h_k he has to use and his only choice is a guess. Due to the Definition 2.62.6 the probability of success is $1/|\mathcal{T}|$. Second possibility for the forger is to wait for a message pair (m, t). With the knowledge of the tuple (m, t), the class of hash function $\mathcal{H}$ and enough computing power, the forger can find all $|\mathcal{H}|/|\mathcal{T}|$ matching keys or hash functions, respectively. Nevertheless, according to Definition 2.62.6 with this knowledge he has a probability of at most ε to guess the correct authentication tag for a modified message m', where $m' \neq m$.

So the aim is to make the ε very small concerning the security aspect. This means to bring it down to $1/|\mathcal{T}|$, which is the lower bound. The disadvantage when $\varepsilon = 1/|\mathcal{T}|$ is that the number of hash functions is very high. Since we need a string (key) to address the hash function, the size of it increases with $\log_2 |\mathcal{H}|$, which results in key length as long as the message length or higher ($|k| \in O(|m|)$). This is not practicable for applications, so we have to find an ε small enough but which increases the key length logarithmic in comparison to the message length.

The first solution for an *almost strongly universal*$_2$ class increasing the key logarithmic was presented by Wegman and Carter [9]. Their construction is shown in Fig. 2.4. Let $s = b + \log_2 \log_2 a$, where $b = \log_2 |\mathcal{T}|$ is the length of an authentication tags and $a = \log_2 |\mathcal{M}|$ is the length of the messages. Construct a *strongly*

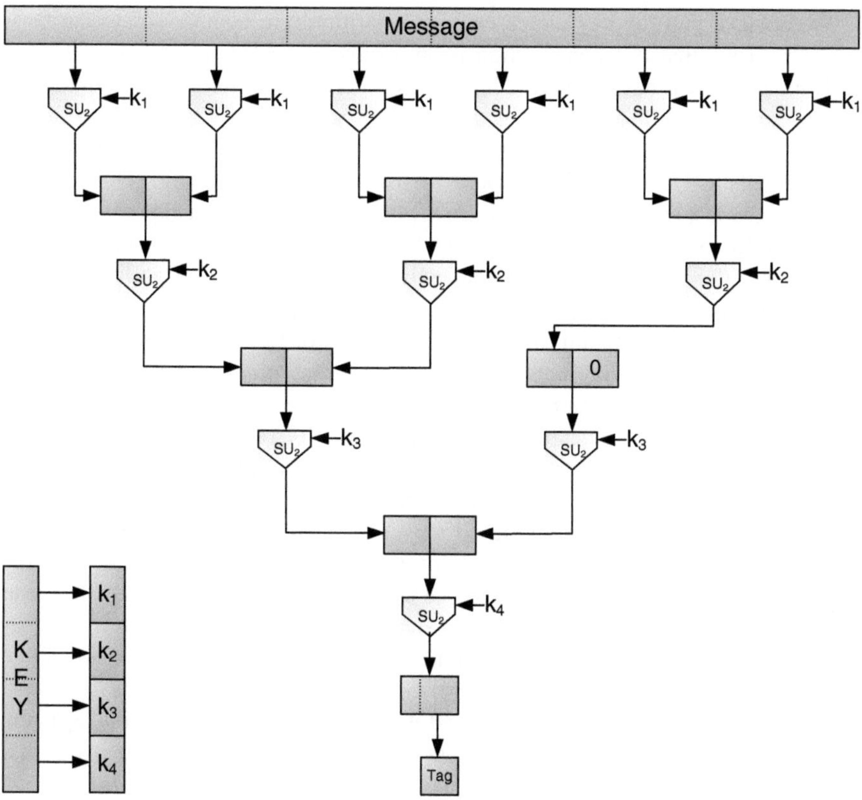

Fig. 2.4 Schematic of $2/|\mathcal{T}|$-*almost strongly universal*$_2$ class of hash function

universal$_2$ class of hash functions, which maps an input string of length $2s$ to a hash value of length s. The idea now is to split the message in blocks of length $2s$. If necessary, add zero padding at the end. Now pick a hash function $h \in \mathcal{H}$ and apply it on all blocks. The iteration ends with concatenating all outputs. Now start the new iteration again with splitting this string in blocks of length $2s$ and picking a hash function until only one block with length s remains. Finally the tag is the low-order $|\mathcal{T}|$ bits of this block. To generate a tag we need $\log_2 a - \log_2 b$ hash functions or keys, respectively. Before we mentioned that the key length in *strongly universal*$_2$ classes increases $O(|m|)$. Now the input for the hash function is a block of length $2s$. A key for one iteration would be $O(2s) = O(s) = O(b + \log_2 \log_2 a)$. We need $\log_2 a - \log_2 b$ iteration resulting in a key length of $O((b + \log_2 \log_2 a) \cdot \log_2 a)$. For further use we will use the suggestions in [9], where a *strongly universal*$_2$ class was used, needing a key roughly twice the size of the input. With this class the following equation is definitively an upper bound for the key size:

$$4 \cdot ((b + \log_2 \log_2 a) \cdot \log_2 a). \tag{2.35}$$

We achieve that the key increases only logarithmic in comparison to the message. To prove that this construction is an *ε-almost strongly universal$_2$* class of hash functions, we have to take a look on the iterations. After every iteration the probability for two distinct messages m_1, m_2 to be equal is $1/2^s$. Since we iterate $\log_2 a - \log_2 b$ times, the probability that the final tags match is at most $\log_2 a/2^s$, which equals to $1/2^b = 1/|\mathcal{T}|$. Because we use a *strongly universal$_2$* class of hash functions the last reduction will be taken m_1 to any tag t_1 with equal probability and fulfills the first part of Definition 2.6a. And as long as the penultimate blocks of m_1 and m_2 are different, m_2 will also be taken into any tag t_2 with probability less than $1/|\mathcal{T}|$, but if these blocks are the same, less than $2/|\mathcal{T}|$ hash functions will take m_2 to any t_2. So ε is $2/|\mathcal{T}|$ and it fulfills the second part of Definition 2.6b.

The second solution presented in [9] tries to reduce the key length needed per message. Therefore again a *strongly universal$_2$* class $\mathcal{H}$ is constructed, which maps $\mathcal{M}$ to $\mathcal{T}$. The two communicating parties (Alice and Bob) split the shared key in two parts. The first part has the length $\log_2 |\mathcal{H}|$ to specify a hash function and the second part is a sequence $(r_1, r_2, ...)$ of elements of $\mathcal{T}$, each with length $\log_2 |\mathcal{T}|$. Secure authentication requires unique indexed messages. The shared secret key indicates both parties which hash function h to choose. To create now the authentication tag t_i for a message m_i with a unique message number (e.g., i), it is hashed by the hash function and then exclusive-or's with r_i, so that $t_i = h(m_i) \oplus r_i$.

For the probability of guessing an authentication tag t for a message m, we define a set for the tag t as $S_t = \{(h, r)|h \in \mathcal{H}, r \in \mathcal{T}, h(m_1) \oplus r = t_1, \text{ and } h(m) \oplus r = t\}$, where m_1 and t_1 are the first message and its authentication tag. In words, S_t is the set of partial keys which map the new message m to the tag t with the knowledge that m_1 is mapped to t_1. Since the chosen class of hash function is *strongly universal$_2$*, the size of all sets S_t is the same. The only way to extend these partial keys of S_t is to append $r_i = h(m_i) \oplus t_i$ such that message m_i maps to tag t_i for $i = 2, 3, ...$ Thus, the sets have the same size and m will be assigned to any tag $t \in \mathcal{T}$ with same probability as any other tag. So the guessing probability of a forger depends on the tag length which results in $1/|\mathcal{T}|$.

Other than the first construction of Wegman and Carter, the needed key size per message depends on the tag length $\log_2 |\mathcal{T}|$ (neglecting the specifier for the hash function). The first construction returns one authentication tag for one large message, in contrast the second construction splits the message in smaller pieces, numbers them, and generate for each piece an authentication tag. In Fig. 2.5 we compare both constructions. Let $\mathcal{H}$ be a *strongly universal$_2$* class of hash functions, where $h : \mathcal{M} \to \mathcal{T}$ and $m_i, m_{j,k} \in \mathcal{M}$, $m' \notin \mathcal{M}$, $t_i \in \mathcal{T}$, $t' \in \mathcal{T}'$, $|m_i| = 2|t_i|$, $i = 1, \ldots, n$, $j = 1, \ldots, |m'|/|t|$, $k = 1, \ldots, \log_2(|m'|/|t|)$ and $lb_b(x) : b$ lower order bits of x. The additional needed space for the numeration of messages m_i is neglected.

The probability to insert a forge tuple (m_i', t_i'), even with knowledge of a tuple (m_i, t_i) is $1/|\mathcal{T}|$. Contrary to the first construction, the probability to insert a forge tuple (m'', t'') even with knowledge of a tuple (m', t') is $2/|\mathcal{T}|$. For the second construction the input size (message size) of the *strongly universal$_2$* class is much larger than the output size (tag size).

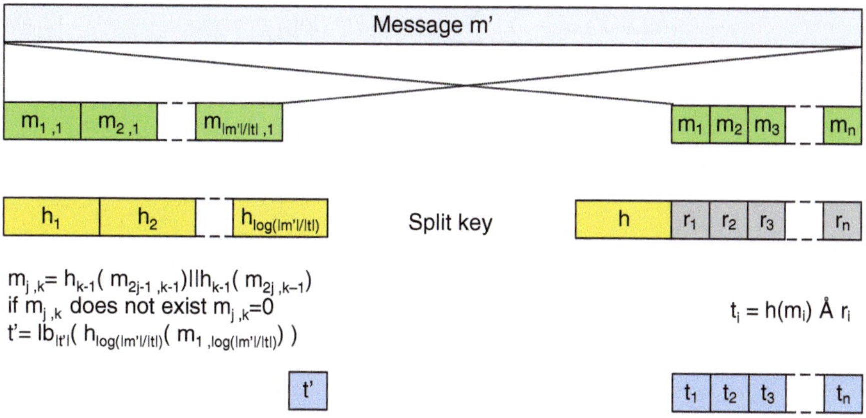

Fig. 2.5 The two constructs of Wegman and Carter [9]

2.3 Entropy

Entropy is a key concept in the field of information theory, which is a branch of applied mathematics concerned with the process to quantifying information. In the following a short part is presented, which we need in later chapters. In general it is recommendable to read up on information theory, hence we refer to the book of Cover and Thomas [2], where the subject is handled in detail.

2.3.1 Shannon Entropy

The Shannon entropy [7] is a basic measure in information theory. Let X be a discrete random variable on a finite set $\mathcal{X} = \{x_1, \ldots, x_n\}$ than the Shannon entropy (or information entropy) $H(X)$ is defined as

$$H(X) = -\sum_{x \in \mathcal{X}} p(x) \log_2 p(x). \tag{2.36}$$

where $p(x)$ is the probability distribution function of X. In other words, $H(X)$ is the expected value of the amount of bits needed to specify a value $x \in \mathcal{X}$. Therefore it is easy to see that the entropy is upper bounded by

$$H(X) \leq \log_2 |\mathcal{X}|, \tag{2.37}$$

with equality if $p(x) = 1/|X|$ for each x. For probabilities $p(x) = 0$ we have the convention $0 \log_2 0 = 0$. Usually the base of the logarithm is 2, so the entropy is measured in *bit*. Other bases are e and 10.

If X and Y are random variables on $\mathcal{X}$ and $\mathcal{Y}$, respectively, the conditional Shannon entropy $H(X|Y)$ is defined as

$$H(X|Y) = -\sum_{y \in \mathcal{Y}} \sum_{x \in \mathcal{X}} p(y)p(x|y) \log_2 p(x|y), \tag{2.38}$$

where $p(x|y) = \frac{p(x,y)}{p(y)}$ is the conditional probability distribution of X given Y.

Further, the mutual information $I(X;Y)$ is defined as

$$I(X;Y) = \sum_{y \in \mathcal{Y}} \sum_{x \in \mathcal{X}} p(x,y) \log_2 \frac{p(x|y)}{p(x)} = \sum_{y \in \mathcal{Y}} \sum_{x \in \mathcal{X}} p(x,y) \log_2 \frac{p(x,y)}{p(x)p(y)}. \tag{2.39}$$

Note that it is called mutual because

$$I(X;Y) = I(Y;X). \tag{2.40}$$

To understand the coherency between $H(X), H(Y), H(X|Y), H(Y|X)$, and $I(X;Y)$ take a look on Fig. 2.6.

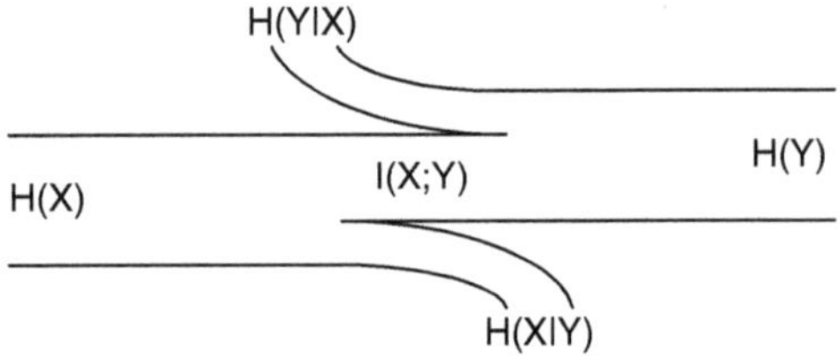

Fig. 2.6 Interpretation of $H(X), H(Y), H(X|Y), H(Y|X)$, and $I(X;Y)$

2.3.2 *Rényi Entropy*

The Rényi entropy [5] of order α is a generalization of Shannon entropy. In the same manner as Shannon entropy, let X be a discrete random variable on a finite set $\mathcal{X} = \{x_1, ..., x_n\}$ than the Rényi entropy of order α $H_\alpha(X)$ with $\alpha \geq 0, \alpha \neq 1$ is defined as

$$H_\alpha(X) = \frac{1}{1-\alpha} \log_2 \sum_{x \in \mathcal{X}} p(x)^\alpha, \tag{2.41}$$

where $p(x)$ is the probability distribution function of X. In the case $\alpha \to 1$, $H_\alpha(X)$ converges to Shannon entropy $H(X)$. Again we use for the logarithm base 2 and the output of the entropy is *bit*. We pick out the Rényi entropy of order 2 ($R(X)$ for short):

$$R(X) = H_2(X) = -\log_2 \sum_{x \in \mathcal{X}} p(x)^2. \tag{2.42}$$

If we consider only the term in the logarithm, we observe that it equals the collision probability. The collision probability $p_c(X)$ of X is defined as the probability that X results in the same value or event twice in two independent executions. The connection to Shannon entropy is that it is upper bounded by Shannon

$$R(X) \leq H(X), \tag{2.43}$$

for every probability distribution $p(x)$.

If X and Y are random variables on $\mathcal{X}$ and $\mathcal{Y}$, respectively, the conditional Rényi entropy $R(X|Y)$ is defined as

$$R(X|Y) = -\sum_{y\in\mathcal{Y}} p(y)\log_2 \sum_{x\in\mathcal{X}} p(x|y)^2, \tag{2.44}$$

where $p(x|y)$ is the conditional probability distribution of X given Y. The upper bound can be derived following Eq. 2.43 as

$$R(X|Y) \leq H(X|Y). \tag{2.45}$$

References

1. Carter, L., Wegman, M.N.: Universal classes of hash functions. J. Comput. Syst. Sci. **18**(2), 143–154 (1979)
2. Cover, T.M., Thomas, J.A.: Elements of Information Theory. Wiley-Interscience (1991). URL http://www.amazon.ca/exec/obidos/redirect?tag=citeulike09–20&path=ASIN/0471062596
3. National Institute of Standards and Technology: FIPS PUB 186–2: Digital Signature Standard (DSS). National Institute for Standards and Technology, Gaithersburg, MD, USA (2000). URL http://www.itl.nist.gov/fipspubs/fip186–2.pdf
4. Nielsen, M.A., Chuang, I.L.: Quantum Computation and Quantum Information. Cambridge University Press, Cambridge (2000). URL http://www.amazon.ca/exec/obidos/redirect?tag=citeulike09–20%&path=ASIN/0521635039
5. Rényi, A.: On measures of information and entropy. Proceedings of the 4th Berkeley Symposium on Mathematics, Statistics and Probability, pp. 547–561 (1961)
6. Rivest, R.L., Shamir, A., Adleman, L.M.: A method for obtaining digital signatures and public-key cryptosystems. Commun. ACM **21**(2), 120–126 (1978)
7. Shannon, C.E.: A mathematical theory of communication. The Bell System Technical Journal **27**, 379–423, 623–656 (1948). URL http://cm.bell-labs.com/cm/ms/what/shannonday/shannon1948.pdf
8. Stinson, D.R.: Universal hashing and authentication codes. In: J. Feigenbaum (ed.) CRYPTO, *Lecture Notes in Computer Science*, Vol. 576, pp. 74–85. Springer, Heidelberg (1991)
9. Wegman, M.N., Carter, L.: New hash functions and their use in authentication and set equality. J. Comput. Syst. Sci. **22**(3), 265–279 (1981)
10. Wootters, W.K., Zurek, W.H.: A single quantum cannot be cloned. Nature **299**(5886), 802–803 (1982). DOI 10.1038/299802a0. URL http://dx.doi.org/10.1038/299802a0

Chapter 3
Quantum Key Distribution

M. Pivk

In this chapter a complete QKD protocol is presented, starting from the transmission via the quantum channel up to the communication over the public channel. The protocol described here is the BB84 protocol, named after Bennett and Brassard [5]. There are other protocols like the B92 protocol [3], the six-state protocol [8], the SARG protocol [19] and the Ekert protocol [10], which are not discussed here. We are focusing on BB84, the most known QKD protocol. In Fig. 3.1 you see an abstract sequence diagram of BB84. k_a is the pre-shared secret needed for authentication and K is the final key generated after BB84 is executed. We begin with the first stage, the transmission of the photons, which is the physical representation of the qubits, from Alice to Bob. This phase of the protocol is discussed in detail in Sect. 3.1. Afterward the communication switches to the public channel (Sect. 3.2).

There, the first phase is the sifting phase, where Alice and Bob negotiate which bits are used and which bits are discarded. To avoid a man-in-the-middle attack by Eve, this message exchange must be authenticated. After agreeing on the bits and being sure that Eve has not modified messages by using an authentication scheme, Alice and Bob go on to the reconciliation phase or error correction phase. Because the quantum channel is not a noiseless channel, Alice and Bob do not share the same identical string. There is a small portion of errors in Bob's string, which is corrected in this phase. Again Eve has the possibility to modify messages during this phase to her interest. Therefore, Alice and Bob must authenticate this phase. Passing reconciliation, Alice and Bob share a string, which is identical with very high probability. But this string cannot be used as a key yet. Eve's information about the string must be considered. She has gained information during the error correction and maybe also during the quantum transmission. Hence, Alice and Bob must map their string via a function to a smaller subset, so that Eve's knowledge decreases almost to zero. This stage is called privacy amplification and afterward Alice and Bob share a secret key only known by them.

M. Pivk (✉)
Safety & Security Department, Quantum Technologies, AIT Austrian Institute of Technology GmbH, Lakeside B01A, 9020 Klagenfurt, Austria, mario.pivk@ait.ac.at: http://www.ait.ac.at

Pivk, M.: *Quantum Key Distribution*. Lect. Notes Phys. **797**, 23–47 (2010)
DOI 10.1007/978-3-642-04831-9_3 © Springer-Verlag Berlin Heidelberg 2010

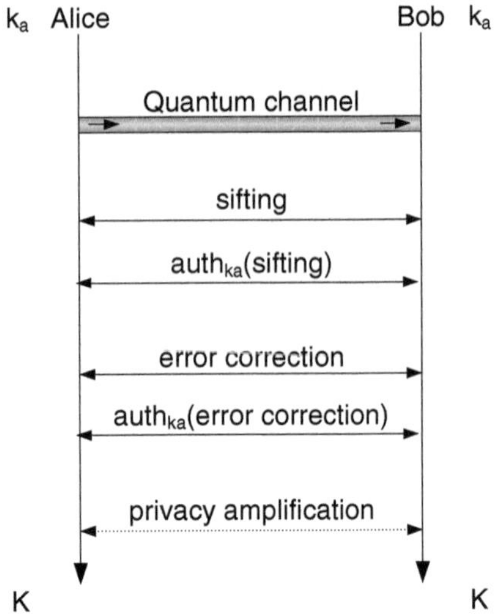

Fig. 3.1 BB84. k_a is the pre-shared key. K is the generated key after protocol execution

3.1 Quantum Channel

As mentioned, the first stage of the BB84 protocol is the quantum transmission. Therefore, we take a closer look at the physical realization: How can qubits be represented and transmitted? In the second part we discuss what Alice and Bob have to do, when they perform the BB84 protocol. Additionally, an equation is given, which represents the theoretical throughput of the quantum channel.

3.1.1 Physical Realization

One way to represent a qubit is by using single photons. They are very suitable for QKD, because photons hardly interact with each other and they can overcome long distances with low loss in optical fibers. Polarization is one of the photon's attributes which can serve as a qubit property. There exists rectilinear polarization (horizontal/0° and vertical/90°) and diagonal polarization (+45° and −45° shifted). We can map horizontal and vertical with the qubits states $|0\rangle$ and $|1\rangle$, and +45° and −45° shifted with the states $|+\rangle$ and $|-\rangle$, respectively. Thus, we have the same statistical probabilities as discussed in Sect. 2.1.3, which we will use next.

There are many different systems for implementing QKD. In Chap. 6 seven systems are presented which are developed for SECOQC.

One possibility to generate single photons is by attenuating the output of a laser. This attenuator is characterized by the mean photon number, which defines the rate of photons passing the attenuator. Usually, the mean photon number is 0.1, meaning that 90% of the time no photon passes the attenuator but in 10% of cases, the probability that a single photon passes is better than 95% (for more information on the computation of the probability see [17]). Nevertheless, the output of the laser may consist of several path; in BB84 due to the amount of different polarization states four are needed. But after a pulse passes the attenuator it has an arbitrary polarization; therefore, linear polarizers and half-plates are used to create the desired polarization. These elements can be seen as linear operators (Pauli matrices) known from Sect. 2.1.2 performing the bit and phase flip on the qubits.

After the photons have the right polarization they pass through a beam splitter, where they are combined. Now the pulse is transmitted over the quantum channel (optical fiber or free space) and arrives at the receiver. There, a random choice is made by the receiver as to which polarization filter to use (e.g., horizontal or +45°). If the receiver chooses differently to the sender we know from Sect. 2.1.3 that there is a chance of 50% for the photon to pass the filter (and change the polarization to horizontal) or to get absorbed (this holds for both +45° and −45° polarized photons). If the receiver chooses the same basis as the sender the photon passes with 100% probability if it was polarized horizontal and with 0% probability if it was polarized vertical. Behind the filter a detector clicks every time a photon passes. (this sending–receiving architecture is taken from [22]).

3.1.2 Photon Transmission and Throughput

Back to our scenario, Alice is on the transmitter and Bob on the receiver site of the above-described quantum channel. Alice chooses now randomly two strings independent of each other with length m. The first string represents the basis for the quantum transmission and the second the proper value of the specific bit. Alice and Bob have the same mapping scheme in common. A sample mapping is defined in Tables 3.1 and 3.2.

Table 3.1 Base mapping

Base	Representation bit
Rectilinear	0
Diagonal	1

Table 3.2 Value mapping

Rectilinear	Diagonal	value bit
Horizontal (0°)	+45°	0
Vertical (90°)	−45°	1

Alice starts now to transmit. Therefore, she takes the first bit of the first string, indicating which base to use (e.g., a 0-bit denotes to use the rectilinear basis), and the first bit of the second string indicating which value to take (e.g. a 1 denotes in the rectilinear base to polarize the photon vertical and in the diagonal base to shift the polarization by −45°). She applies this procedure on all m-bits of both strings and sends them as photons via the quantum channel to Bob. Bob on his part chooses also a random string with length m, for his base choices. With these bits he measures the incoming photons with the corresponding filter. Note that it makes no sense to measure with both filters, because after measurement the original polarization is lost as mentioned in the previous section or in Sect. 2.1.3. Bob keeps the measurement results to himself. Consider that Bob's measurements do not completely match with Alice's ones, due to optical misalignment, disturbance on the quantum channel, noise in Bob's detectors, or the presence of an eavesdropper Eve, although both choose the same basis.

Before describing the next step, we continue by analyzing the throughput of the quantum channel. In [24] a theoretical function is given by which the several losses on the quantum channel can be computed:

$$g_q = \mu \cdot \alpha_f \cdot \alpha_e \cdot \eta_{\text{det}} \cdot k_{\text{dead}}. \tag{3.1}$$

The gain of the quantum channel g_q consists now of the mean photon number μ, which we discussed in the previous section. The next factor α_f represents the fiber loss. This loss is distance dependent and increases with higher distances. For the receiver's detector η_{det} as detection efficiency and k_{dead} as factor accounting for the reduction of the photon detection rate due to the dead time is given. The dead time is the hold-off time following each detection event; during this time the bias voltage of the device is below a certain level such that no photon can be detected. α_e is the additional loss of the system.

To compute the number of Bob's measurement results the data rate f_{data} is required, which is the number of photons Alice's laser can send. This rate is given in hertz and can have range from MHz to GHz. These different rates have impact on the gain of the quantum channel: with higher rates the efficiency shrinks. So the key material Bob receives per second can be computed by $f_{\text{data}} \cdot g_q$. The Eq. 3.1 conforms with experimental test done in [24].

As long as the strings which Alice and Bob need for the choice of bases are randomly chosen (the probability to send a qubit in the rectilinear base or diagonal base is 50%), half of Bob's results, i.e., the raw key, must be discarded because of the independence of both strings (see later in Sect. 3.2.1). To minimize this loss of raw key bits a concept was designed in [13], which increases the efficiency. Instead of having the same probability for 0's and 1's in the string, the authors propose to increase the probability for one of them. In fact the photons will be transmitted more often in one base than in the other. Thus, the rate in which Alice and Bob select the same base will increase. We will discuss this more precisely in Sect. 3.2.1.

Finally, the communication between Alice and Bob on the quantum channel is finished. They now switch to the public channel. So far, Alice holds the two strings of length m containing her choice of base and value and Bob a string of length m containing his choice of bases and his measurement results of length $g_q \cdot m$. The smaller size is due to the loss on the quantum channel.

3.2 Public Channel

The communication via the public channel is necessary for Alice and Bob, since they must negotiate on which bits they perform the next steps. Because they choose their bases for the quantum transmission randomly they must know in which cases they used the same basis. After agreeing on them, the errors must be corrected, since the communication via the quantum channel is noisy. And as a last step Alice and Bob must reduce Eve's knowledge, which she gained during the protocol. By intercepting and resending photons Eve gains knowledge of the raw key (this issue is reflected by the error rate) and also eavesdropping on the messages on the public channel increases her knowledge. If Eve has the possibility to modify messages, she can start a man-in-the-middle attack and then share keys with both disguised as one's peer.

3.2.1 *Sifting*

The first phase on the public channel is *sifting*. After Alice has sent random bits mapped into randomly chosen quantum bases via photons, further steps are required such that she shares the same bit string with Bob. The first thing Alice must know is which photons Bob has measured. By reason of the fact that only a small fraction of pulses contain photons, and several losses on the quantum channel and at Bob's detector, Alice does not know which bit Bob received. Therefore, Bob sends a message on the public channel to inform Alice which photons he has measured. The easiest way is to send a string as long as the string chosen at the beginning for the bases, which we indicate with length m. Bob sets now the position where he was able to make a measurement to 1 and the other positions to 0. Since we want to save communication traffic, we can benefit from the fact that the gain of the quantum channel g_q (Eq. 3.1) is a very small factor (<0.01). Bob must only tell Alice which position he received. To represent a position $\log_2 m$-bits are necessary. Let $2n$ be the length of the raw key, which is equal to $g_q \cdot m$, then $2n \log_2 m$-bits are necessary to tell Alice the measured positions. This would be more efficient if the following equation is fulfilled:

$$2n \log_2 m = g_q \cdot m \log_2 m < m.$$

We know that the gain of the quantum channel is less than 0.01, so it is sufficient that $m < 2^{100}$. The sending rate is maximal in the range of GHz ($10^9 \approx 2^{30}$), hence adequate for this sifting transmission mode.

With this message Alice knows only which position Bob has measured but not if he has used the same basis, because they have randomly and independently chosen the bases. Thus, Bob sends a second message, in which he publishes the bases he used. An optimized approach sends only those positions where he has successfully measured a photon. Thus, the length of the second message is $2n$. Note that Bob sends only his sequence bases (see Sect. 3.1.2) reduced by the bits where no measurement was possible, and not the measurement results. After Alice receives both messages she can reduce her secret bit value string by canceling out those positions which Bob did not receive (Bob's first message) and those positions where she uses different bases at transmission (comparison by her base string and Bob's second message). To make sure Bob shares the same string with her. Alice sends a message to Bob containing her choice of bases for those positions where Bob received a photon. Bob himself performs the same procedure as Alice and cancels out those positions with different bases. Finally, the sifting phase is over and Alice and Bob share a secret key: the so-called sifted key. However, error remains due to the noisy quantum channel.

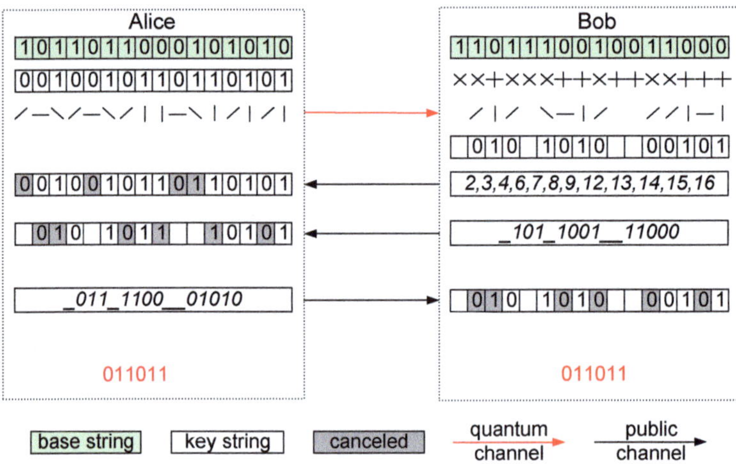

Fig. 3.2 QKD protocol until the end of the sifting phase

But what is the length of this key? Since Alice and Bob have chosen the two bases for transmission and measuring randomly and independently, the probability that they use the same basis is $\frac{1}{2} \cdot \frac{1}{2} + \frac{1}{2} \cdot \frac{1}{2} = \frac{1}{2}$. The length of the sifted key reduced by mismatching bases from the raw key is $\frac{2n}{2} = n$. The complete process until the output of the sifted key is demonstrated in an example in Fig. 3.2. The sending key length is $m = 16$. Assuming that the gain of the quantum channel is given by $g_q = 0.75$ (which is not a realistic value), we use Tables 3.1 and 3.2 for mapping bits to photons. So the sifted key length is $n = m \cdot g_p \cdot \frac{1}{2} = 6$.

As mentioned in Sect. 3.1.2 it is suggested [13] to choose the strings representing the photon bases with different probability. So far, we chose the rectilinear or the diagonal basis with probability $\frac{1}{2}$. Assume now that Alice and Bob choose a basis with probability $\delta \in [0, 1]$, whereby the other basis is used with complementary probability $1-\delta$. The protocol gain for the normal BB84 is 0.5 as we have computed before; the protocol gain g_p for the case that the probability of base usages can be changed is

$$g_p = \delta \cdot \delta + (1 - \delta) \cdot (1 - \delta) = 2\delta^2 - 2\delta + 1. \tag{3.2}$$

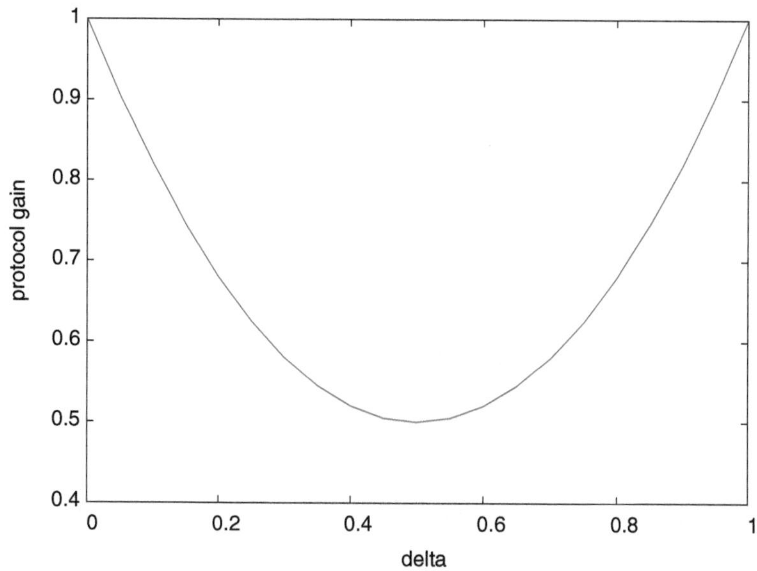

Fig. 3.3 Protocol gain g_p when delta δ is variable (Eq. 3.2)

The fact that δ is variable gives us the probability to increase the protocol gain. As you can see in Fig. 3.3 the gain g_p rises to 1 if δ goes to 1 (to 0 – analog for the other basis). So the case $\delta = 0.5$ would be the worst case.

But what happens with the security if δ goes to 1. In Sect. 3.2.3.1 we will discuss this in detail. To say it right away, the chance for Eve to eavesdrop the photons without being detected increases because Alice and Bob must negotiate at least which bases to use with higher probability, and Eve can use this information. The errors Eve produces decrease in the set where the basis is chosen with higher probability and increase in the set of the other basis. Using the naive error estimation, which calculates the error probability of both sets together, Alice and Bob will fail to detect Eve, because the proportion of the sets changes. Finally, when δ reaches 1, Eve will not introduce any errors. Unless $\delta = 1$, if we use a refined error analysis, which reflects the error probability of both sets (detailed explanation in Sect. 3.2.3.1), Eve can still be detected. In [13] it is described that the set of bits after sifting must have enough samples to make an accurate error estimation. Therefore, some constraints on δ are defined. Assume that m_t photons are transmitted from Alice to Bob, which Bob was able to measure. Then, on average, only $m_t(1-\delta)^2$ photons belong to the case where the basis with the smaller probability is used. The size of this set was analyzed in [13] with the information we allow Eve to obtain on the final key. If we except a fixed but arbitrarily small amount of information on the final key, the number of this set is required only to scale logarithmically with the length of the final key k. Thus, δ must fulfill the following equation:

$$m_t(1-\delta)^2 \geq O(\log k),$$

$$\delta \leq 1 - O\left(\sqrt{\frac{\log k}{m_t}}\right),$$

$$\text{if } m_t \rightarrow \infty \text{ then } \delta \rightarrow 1 \text{ and } g_p \rightarrow 1. \tag{3.3}$$

Note that δ never reaches 1 and the fact that the protocol gain goes to 1 can easily be seen in Fig. 3.3.

3.2.2 Authentication of Sifting

Due to the possibility of a man-in-the-middle attack (combination of intercept/resend attack, see Sect. 5.2.1, and modification of messages on the public channel) by Eve we must ensure that the sifting phase is authenticated by Alice and Bob. We know this phase consists of three messages exchanged between Alice and Bob. Two messages sent by Bob, where he tells Alice which photons he was able to measure and which bases he used for the measurement. The last message completing sifting phase is sent by Alice, where she tells Bob which bases she used for those photons Bob measured. To authenticate those messages we use Wegman–Carter authentication, as explained in Sect. 2.2.2. For Bob's two messages and for Alice's message we generate authentication tags and append them to the respective messages.

Because Wegman–Carter authentication is symmetric, we have to use a secret key shared by Alice and Bob. Since key material is generated in every iteration a small part of it can be used for the authentication of the next round. Unfortunately, when we start the first round a pre-shared key must be available, which is exchanged previously through a secret channel (face-to-face or in another way). In order that key material remains after withdrawal by authentication, the choice of the algorithm is a major concern. Following Eq. 2.35 the authentication cost for Wegman–Carter is $4 \cdot ((b + \log_2 \log_2 a) \cdot \log_2 a)$ for an input size of a and an authentication tag size of b. The key size grows only in the logarithmic scale, so it would be useful for our intended purpose.

Similar to [11] we compute now the authentication cost for the sifting phase. In the previous section we derived that the length of Bob's first message is $2n \log_2 m$ and the length of the second message is $2n$. The sum is $2n(1 + \log_2 m)$ and thus the authentication key length w_1 needed for the tag is

$$w_1 = 4 \cdot \left(g_{\text{auth}} + \log_2 \log_2 \left(2n\left(1 + \log_2 m\right)\right)\right) \cdot \log_2 \left(2n\left(1 + \log_2 m\right)\right), \tag{3.4}$$

where g_{auth} is the length of the resulting tag. Alice and Bob compute their tag using the hash function indexed by the authentication key and Alice compares if the tags match.

If Alice verifies that the message is from Bob she sends her message with her choice of bases, which has length $2n$. We denote the authentication key length for Alice's message as w_2 and the length is computed as

$$w_2 = 4 \cdot \left(g_{\text{auth}} + \log_2 \log_2(2n)\right) \cdot \log_2(2n). \tag{3.5}$$

If Bob determines that the generated tag with cost w_2 matches with Alice's tag both can be sure that the probability that Eve modified the message is at most ε, which is the property of the class of hash function (ε-ASU_2).

The required key material to authenticate the sifting phase is the sum of all costs w_1, w_2 (Eqs. 3.4 and 3.5) and results in the total sifting authentication costs

$$\begin{aligned} t_S =& 4 \cdot \left(g_{\text{auth}} + \log_2 \log_2 \left(2n \left(1 + \log_2 m\right)\right)\right) \cdot \log_2 \left(2n \left(1 + \log_2 m\right)\right) \\ &+ 4 \cdot \left(g_{\text{auth}} + \log_2 \log_2(2n)\right) \cdot \log_2(2n). \end{aligned} \tag{3.6}$$

3.2.3 Reconciliation

The reconciliation stage is split into two parts. The first major part is the error estimation. Its aim is to find the correct error rate. To guarantee an optimal error correction, part two of reconciliation. The error correction has the task of correcting all errors in Bob's string via public discussion. Here some information about the string must be disclosed and exchanged between Alice and Bob.

3.2.3.1 Error Estimation

The error estimation is an important step in the QKD protocol to determine the proper error rate of the sifted key. Usually, in BB84, the error rate p is estimated by picking a small random subset of bits with length r from those given in the sifted key. This test string is publicly compared by Alice and Bob and yields in a certain number of errors e. If the length of the test string is chosen adequate to the length of the sifted key n, the error probability is

$$p = \frac{e}{r}. \tag{3.7}$$

Of course, since a part of the sifted key has been announced, those bits must be deleted to avoid information leakage to Eve. This elimination has little effect on the length of the final key if the length of the sifted key is large and if the error estimation is not executed every round. The error rate found for a round can also be used in the following rounds.

If the error rate p turns out to be very large, then either eavesdropping has occurred or the channel is somehow unusually noisy. However, the sifted key is discarded and Alice and Bob may re-start the whole protocol again (on another quantum channel). This threshold p_{max}, for which the rate should not exceed $p \leq p_{\text{max}}$, can be set to, e.g., 11% because at the moment the best error correction code approaches a maximal tolerated error rate of 12.9% [21]. If the error rate p is reasonably small ($p \leq p_{\max}$), Alice and Bob can continue with error correction.

An improved error estimation was presented in [1] and [13]. In the refined error analysis there does not compute a single error rate but two values are estimated. The idea is not to merge the measurements during the quantum transmission, where Alice and Bob have used the same basis, into one set (more precisely the sifted key) and choose the random test subset out of it. They choose two random test subsets. The first subset consists of all measurements where Alice and Bob have used the first basis (e.g., rectilinear) and the second one where they used the second basis (e.g., diagonal). Let r_1 be the length of the first subset and r_2 be the length of the second subset just before Alice and Bob publicly compare both strings and estimate the number of errors e_1 and e_2, respectively, which leads to the error rates

$$p_1 = \frac{e_1}{r_1}, \tag{3.8}$$

$$p_2 = \frac{e_2}{r_2}, \tag{3.9}$$

where p_1 is the error rate for the photons measured with the first basis and p_2 those with the second basis and

$$p = \frac{p_1 + p_2}{2}. \tag{3.10}$$

The advantage compared to the single error rates is as follows: If Eve starts a specific attack like the biased eavesdropping strategy (a specialization of intercept and resend, see Sect. 5.2.1). Then she chooses the probability q_1 for measuring each photon sent from Alice to Bob in the first basis (e.g., rectilinear) and a second probability q_2 for measuring each photon in the second basis (e.g., diagonal). Hence, with probability $1 - q_1 - q_2$ she does not measure the photon. When Alice and Bob use the same basis errors occur only if Eve uses a different basis. Regarding Bob's side, the photons measured by Eve are randomized and yield an incorrect bit in half of these cases. This introduces an error rate of $p_1 = \frac{q_2}{2}$ for the first basis. Similarly, for the second basis an error rate $p_1 = \frac{q_2}{2}$ is obtained (without respect to the normal noise on the quantum channel). If we apply these two error rates and Eq. 3.10 to the requirement $p \leq p_{\max}$ it results in $(q_1 + q_2) \leq 4p_{\max}$. Eve has the possibility to vary her probability q_1 in a big range. In contrast if we use the constraint $p_1, p_2 \leq p_{\max}$ which has the same property as the single error rate constraint in a random noisy channel, Eve's possibility to choose the probabilities is shortened with $q_1, q_2 \leq 2p_{\max}$. With the refined error analysis the biased eavesdropping strategy is avoided (e.g., $q_1 = 3p_{\max}$, $q_2 = 0$ is accepted when we measure only one error rate, but rejected by the detailed error measurement).

In Sect. 3.2.1 we have heard about a strategy to vary the probability of used bases at sending and receiving, to boost the protocol gain g_p (in standard BB84 protocol the probability for sending in one of the bases is $\frac{1}{2}$ as described in Eq. 3.2).

The difficulty now is that with increasing or decreasing δ, a biased eavesdropping strategy where Eve aware of δ is more difficult to detect using the simple error estimation rate. In this case the simple error rate is

$$p = \frac{\delta^2 p_1 + (1-\delta)^2 p_2}{\delta^2 + (1-\delta)^2} = \frac{\delta^2 q_1 + (1-\delta)^2 q_2}{2(\delta^2 + (1-\delta)^2)}. \tag{3.11}$$

If Eve chooses the strategy $q_1 = \delta$ and $q_2 = 1 - \delta$ we see in Fig. 3.4 that when $\delta \to 1$ or $\delta \to 0$ the error probability $p \to 0$ unlike the refined error analysis where $\delta \to 1$ then $p_1 \to 0$ but $p_2 \to 0.5$. Regardless of how δ is chosen $(0 < \delta < 1)$ if the measurement set is quite big (as defined in Sect. 3.2.1) then there would be no problem to detect Eve, which is not possible with the simple error analysis.

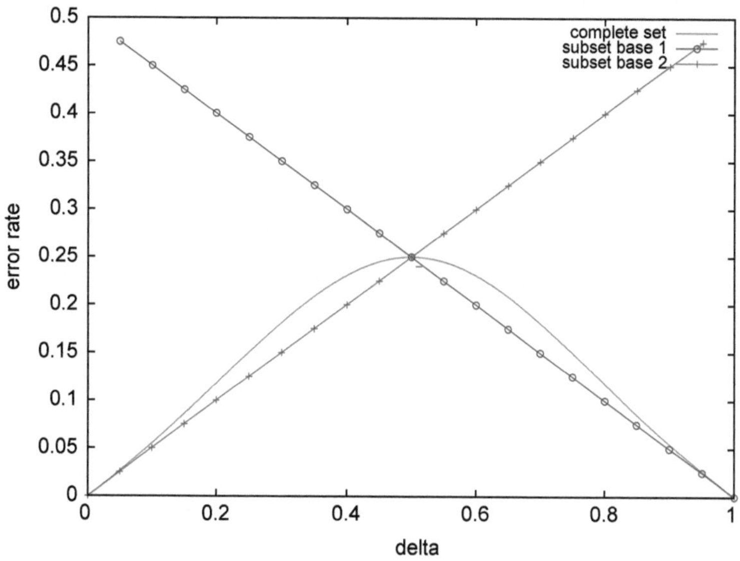

Fig. 3.4 Error rates (variable δ) after a biased eavesdropping attack with $q_1 = \delta$

3.2.3.2 Error Correction

Henceforth we will model the quantum channel as binary symmetric channel (*BSC*). In a $BSC(p)$ we have an alphabet of two symbols (0, 1) which are transmitted over the channel from a sender to a receiver. With probability p noise is added to the transmitted symbol, meaning the receiver gets a 1 instead of a 0 and a 0 instead of a 1, respectively. For every symbol we have a probability $1-p$ that it will be correctly received (see Fig. 3.5).

After the sifting phase Alice holds a random string sk_a and Bob sk_b with equal length $|sk_a| = |sk_b| = n$. The difference between these two strings depends on the quantum bit error rate (*QBER*) p of the channel. Starting the QKD protocol

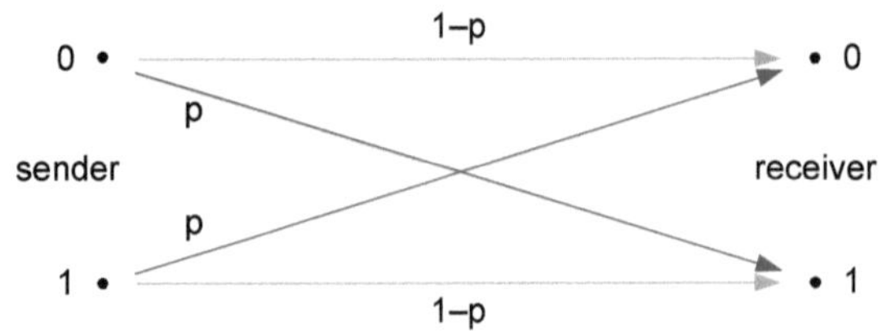

Fig. 3.5 Binary symmetric channel (BSC): symbols are exposed to noise with probability p

this probability p is unknown or rests on empirical values. Further on, the error estimation phase as presented in Sect. 3.2.3.1 gives a good approximation of the error rate. Observe that dist (sk_a, sk_b), the amount of places in which both strings differ (Hamming distance), is nearly np. Setting $sk_a = A$ and $sk_b = B$ (A and B are random variables) the conditional entropy (Eq. 2.38) of A given B is

$$H(A|B) = H(A \oplus B) = nh(p),$$

where $h(p)$ is now short for Shannon entropy $H(X)$ and the random variable X is a Bernoulli trial with parameter p (for each $x \in \mathcal{X}$, $p(x)$ is the same).

The task is to correct Bob's string B without disclosing enough information that gives Eve chances to reconstruct the very same string. Therefore, a reconciliation protocol R^p is defined, which runs on both strings A, B and results in the string S by exchanging some information Q on the public channel. We write $S = \perp$ if the protocol fails to produce S. Since Q must be exchanged on the public channel, an eavesdropper Eve can gain some information on S, which can be expressed by $I_E(S|Q)$. This is the expected amount of bits that an eavesdropper Eve can get on S given Q.

Brassard and Salvail [7] have formulated some definitions to characterize such reconciliation protocols. The first definition deals with the robustness.

Definition 3.1 A reconciliation protocol R^p is ε-robust if

$$(\exists N_0(\varepsilon))(\forall n \geq N_0(\varepsilon)) \sum_{\alpha,\beta \in \{0,1\}^n} prob(A = \alpha, B = \beta)\; prob(R^p(\alpha, \beta) = [\perp, \cdot]) \leq \varepsilon$$

where $0 \leq \varepsilon \leq 1$.

If a protocol is ε-robust the probability to fail is maximal ε. The next theorem is a direct consequence of the noiseless coding theorem.

Theorem 3.1 $(\forall p \leq \frac{1}{2})$ *($\forall$reconciliation protocol R^p) If there exists $0 \leq \varepsilon \leq 1$ such that $R^p = [S, Q]$ is ε-robust then*

$$\lim_{n \to \infty} \frac{I_E(S|Q)}{nh(p)} \geq 1,$$

where n is the length of the transmitted string.

In other words, Eve's information of S is greater than or equal to the information of A given B. Eve cannot have less than this information if the reconciliation protocol R^p is successful. But if these two pieces of information are equal the protocol is optimal. Let us define it by the next definition:

Definition 3.2 A protocol R^p is optimal if

$$(\forall \varepsilon \geq 0)[R^p = [S, Q] \text{ is } \varepsilon\text{-}robust]$$

and

$$\lim_{n \to \infty} \frac{I_E(S|Q)}{nh(p)} = 1,$$

where the public channel is a *BSC(p)*.

In [7] a construction of a optimal protocol is presented. The basic idea is that Alice creates a random label of her string A with length approximately $nh(p)$, when the length of string A is n. If Alice sends this label over the channel, Eve would only have this information and so Definition 3.2 holds, the protocol is optimal. When Bob obtains the label $f(A)$, he computes all possible inputs, resulting in a set $\mathcal{S}\{B'|f(B') = f(A)\}$. The string $B' \in \mathcal{S}$ with minimal Hamming distance from B is the desired S. Brassard and Salvail [7] described this protocol as unpractical, because Alice and Bob require 2^{m2^n} functions. But if the function $f(x)$ is chosen from a $universal_2$ class of hash functions (see Definition 2.5) the protocol remains optimal. The specification of a $universal_2$ class of hash functions can be done in a short and efficient way. The only problem which remains is that there are no known efficient algorithms to compute the set $\mathcal{S}\{B'|h(B') = h(A)\}, h \in \mathcal{H}$, if the hash value of A is known.

The goal is to create an efficient reconciliation protocol, which is defined as follows:

Definition 3.3 A reconciliation protocol R^p is efficient if there is a polynomial $t(n)$ such that $\bar{T}^{R^p}(n) \leq t(n)$ for n sufficiently large, where n is the length of the strings transmitted over the secret channel and $\bar{T}^{R^p}(n)$ represent the expected running time of R^p, given an n-bit long input string.

Definition 3.4 A reconciliation protocol R^p is ideal if it is both optimal and efficient.

Brassard and Salvail determined when their optimal protocol becomes ideal, using the $universal_2$ class of hash function H_3 [9]. They proved that their protocol is ideal if and only if *NP* $\subseteq$ *BPP*, which is a hypothesis.

With the fact that a ideal reconciliation protocol depends on a open question in complexity, which is unlikely to be true, we have to find another solution. In Sect. 2.2.1 (universal hashing) we had a similar problem. There we define a $strongly$ $universal_2$ class of hash functions, where the size of this class becomes too large and unpractical. But just a small change of the probability on the theoretical bound makes these classes useful. Here again if we do not demand optimality and allow the reconciliation protocol to transmit a small amount of leaked information above the theoretical bound the protocol becomes efficient and useful. In [7] this characterization was defined as in Definition 3.5.

Definition 3.5 A reconciliation protocol R_ζ^p is almost ideal if for all $\zeta \geq 0$ we have

1. $(\exists \varepsilon \geq 0)[R_\zeta^p = [S, Q]$ is ε-robust$]$
2. $\lim_{n\to\infty} \frac{I_E(S|Q)}{nh(p)} \leq 1 + \zeta$
3. $(\exists$ polynomial $t)(\exists N_0(\zeta))(\forall n \geq N_0(\zeta))[\bar{T}^{R_\zeta^p} \leq t(n)]$

for n the length of the string transmitted over the $BSC(p)$.

An *almost-ideal* protocol has an error probability bounded by ε approaching 0 for increasing n (see Definition 3.51). The amount of leaked information is allowed to be slightly greater than the theoretical bound (see Definition 3.52), but the parameter ζ indicating the excess of information is chosen by Alice and Bob before the start of the protocol. After their choice the expected runtime $\bar{T}^{R_\zeta^p}$ of the reconciliation protocol must be bounded by a polynomial (see Definition 3.53).

Such an *almost-ideal* protocol was presented in [7] and an earlier version in [4]. This simple protocol called CASCADE leaks an amount of information close to the theoretical bound, when the error probability is below 15%.

The appropriate part of CASCADE for the error correction is BINARY. When Bob's string contains an odd number of errors, Alice and Bob can perform an interactive binary search on the strings to find and correct one error, respectively. At the beginning Alice sends Bob the parity of the entire string and Bob checks if his parity bit differs from Alice's bit. If not, there is an even number of errors (possibly zero) in the string and nothing is done. If the parity bits differ the following steps are run through to find the error:

1. Alice sends Bob the parity of the first half of the string.
2. Bob determines whether an odd number of errors occurred in the first or in the second half by testing the parity of his string and comparing it to the parity sent by Alice.
3. With the half determined in step 2 we start again at step 1, until the erroneous bit is found.

For a better understanding see the visualization in Fig. 3.6. The leaked information for a string with an even number of errors is one bit and for a string with an odd number $\lceil \log n \rceil + 1$ bits, whereas one error is corrected.

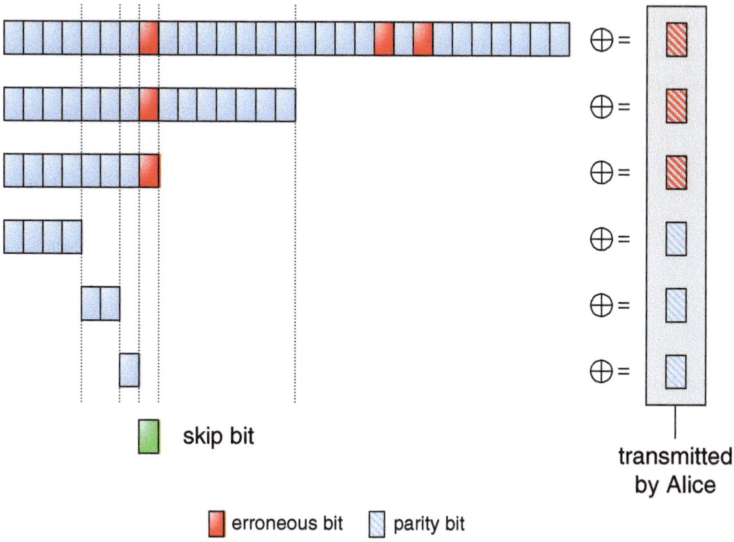

Fig. 3.6 BINARY corrects exact one error of a block with an odd number of errors

CASCADE proceeds in several passes, which are defined by Alice and Bob before execution relative to the error probability p. Alice and Bob hold string $A = A_1, ..., A_n$ and $B = B_1, ..., B_n$ ($A_i, B_i \in \{0, 1\}$), respectively. For the first pass they choose k_1 (determination of parameter k_1 is shown later) and split their strings into blocks of length k_1. Block v for pass 1 is defined by $K_v^1 = \{l|(v-1)k_1 < l \leq vk_1\}$, $v = 1...\left\lceil \frac{n}{k_1} \right\rceil$. On each block BINARY is performed. This can be done parallel for each block to minimize the communication effort. For passes $i > 1$, Alice and Bob choose a k_i and a random function $f_i : [1..n] \rightarrow \left[1..\left\lceil \frac{n}{k_1} \right\rceil\right]$. Now the block j in pass i has the form $K_j^i = \{l|f_i(l) = j\}$. Again Alice and Bob perform BINARY on these blocks to correct errors. In the previous version of CASCADE [4] the leaked information about the string is eliminated during execution of BINARY by removing the last bit of each subset for which a parity bit is computed. In [7] an improvement is introduced where the removed bits are kept. This allows us to correct more errors in later passes. Because if in pass $i > 1$ an error $B_l \neq A_l$ of block K_j^i is corrected, each block K_v^u from previous passes containing this bit $l \in K_v^u$, $1 \leq u < i$, has now an odd number of errors and can be corrected with less effort. Therefore, a set $\mathcal{K}$ of these blocks is created. The smallest block in $\mathcal{K}$ is chosen and BINARY is executed on it. After correcting the error $B_{l'}$, we create again a new set $\mathcal{B}$ with the previous blocks containing bit $B_{l'}$ from each pass from 1 to i. Now set $\mathcal{K}' = \mathcal{B} \cup \mathcal{K} \backslash \mathcal{B} \cap \mathcal{K}$ is determined. If $\mathcal{K}' \neq \emptyset$ then by choosing again the smallest block further errors can be eliminated. Pass i ends when all blocks K_j^i are checked and when there are no more blocks in $\mathcal{K}'$.

For determining the parameter k_i let $\delta_i(j)$ be the probability that after the pass $i \geq 1$, $2j$ errors remain in block K_v^1. Let E_i be the expected number of errors in K_v^1 after pass i. So that after the first pass we have

$$\delta_1(j) = \text{prob}(X = 2j) + \text{prob}(X = 2j+1),\ (X \approx Bin(k_1, p)),$$

$$E_1 = 2 \sum_{j=1}^{\left\lfloor \frac{k_1}{2} \right\rfloor} j\delta_1(j).$$

Suppose now that k_1 is chosen such that

$$\sum_{l=j+1}^{\left\lfloor \frac{k_1}{2} \right\rfloor} \delta_1(l) \leq \frac{1}{4}\delta_1(j) \tag{3.12}$$

and

$$E_1 \leq -\frac{\ln \frac{1}{2}}{2} \tag{3.13}$$

are satisfied and for the following passes we simply define $k_i = 2k_{i-1}$ for $i > 1$. Brassard and Salvail have shown in [7] that if k_1 satisfies Eqs. 3.12 and 3.13 the amount of information $I(\omega)$ per block of length k_1 leaked after ω passes can be bounded as follows:

$$I(\omega) \leq 2 + \frac{1-(1-2p)^{k_1}}{2}\lceil \log k_1 \rceil + 2\sum_{l=2}^{\omega}\sum_{j=1}^{\left\lfloor \frac{k_1}{2} \right\rfloor} \frac{j\delta_1(j)}{2^{l-1}}\lceil \log k_1 \rceil. \tag{3.14}$$

For different error probabilities $p \in \{0.01, 0.05, 0.10, 0.15\}$ in four passes the theoretical bound and the leaked information $I(4)$ are computed with Eq. 3.14. In addition empirical tests were performed to find the average amount of leaked information $\hat{I}(4)$ (10 tests with $n = 10,000$). The results of [7] are present in Table 3.3. The column $f(p)\frac{I(4)}{kh(p)}$ gives the percentage of additional leaked information relative to the theoretical bound.

Table 3.3 CASCADE benchmark

p	k_1	$\hat{I}(4)$	$kh(p)$	$I(4)$	$f(p) = \frac{I(4)}{kh(p)}$
0.01	73	6.47	5.89	6.81	1.16
0.05	14	4.60	4.01	4.64	1.16
0.10	7	3.81	3.28	3.99	1.22
0.15	5	3.80	3.05	4.12	1.35

Finally, the reconciliation phase is finished. After getting random strings for Alice and Bob from the sifting phase with length n and error probability p the described reconciliation protocol (CASCADE) has corrected Bob's string and only a very low probability remains that Bob's string has still an error. The leaked information was kept down near the theoretical bound with an additional leakage of 16–35% depending on the error probability. In Chap. 4 a advancement of this protocol is presented, called Adaptive Cascade.

3.2.4 Confirmation/Authentication of Error Correction

The next step in the protocol is on the one hand to confirm the strings in Alice's and Bob's possession and on the other hand to authenticate the reconciliation phase. Fortunately, this can be done in one step based on the fact that Eve could mount a man-in-the-middle attack during the reconciliation phase. Without the authentication step, Alice and Bob could believe that their strings, on which they operate, are identical when they are not. Only one different bit is necessary to produce completely uncorrelated strings at the privacy amplification phase (see Sect. 3.2.5). The best solution for Alice and Bob is to verify that the outputs of the error correction phase are the same. This will authenticate the complete communication during the reconciliation phase, because an intervention of Eve will introduce errors.

The verification can be done by hashing the corrected random strings and compare the resulting tags. We also should keep the size of the tag small compared to the input (corrected shifted key) size. Every additional bit will increase the leaked information and accordingly Eve's knowledge.

As in Sect. 3.2.2 we have to sacrifice a part of already shared key for authentication. Again the best choice to authenticate the corrected shifted key is to use Wegman and Carter authentication as presented in Sect. 2.2.2. Gilbert and Hamrick acknowledge that this is the best choice and analyzed the total authentication costs for the error correction phase [11]. After Eq. 2.35 the authentication cost is $4 \cdot ((b + \log_2 \log_2 a) \cdot \log_2 a)$ for an input size of a and an authentication tag size of b.

The size of the corrected sifted key at the end of the error correction phase is n. The required authentication key length w_1 for the generation of an authentication tag is

$$w_1 = 4 \cdot (g_{EC} + \log_2 \log_2 n) \cdot \log_2 n, \tag{3.15}$$

where g_{EC} is the length of the resulting tag. Alice and Bob compute the tag using the hash function indexed by the authentication key and compare if the tags match. The main problem is if Bob sends this tag to Alice, Eve can modify the tag with arbitrary bits to convince Alice that the strings do not match when, in fact, they do. To avoid this denial-of-service attack Bob must authenticate his message. Therefore, he computes another tag, which authenticates the previous tag as his tag and sends it to Alice. Again he needs a part of the shared key to generate the tag with length g_{auth} which costs

$$w_2 = 4 \cdot (g_{\text{auth}} + \log_2 \log_2 g_{EC}) \cdot \log_2 g_{EC}. \tag{3.16}$$

If Alice determines that her generated tag with cost w_1 matches with Bob's tag and the authentication tag with cost w_2 does as well, she can be sure that the strings are equal. If the tag with cost w_1 does not match but the second does, there must be at least an error in the string of Bob. In the case that both tags do not match, she can assume that either Eve manipulates the messages or the shared keys of Alice and Bob are different.

To indicate Bob that the tags match Alice sends a piece of the key with length

$$w_3 = \tilde{g}_{EC} \tag{3.17}$$

to Bob. Again to avoid the man-in-the-middle attack, Alice must authenticate her message. With the same tag length g_{auth} as Bob's before, yielding an authentication cost of

$$w_4 = 4 \cdot (g_{\text{auth}} + \log_2 \log_2 \tilde{g}_{EC}) \cdot \log_2 \tilde{g}_{EC}. \tag{3.18}$$

After Bob has compared the tags of Alice with his ones they agree the authentication step for the error correction is complete. Alice and Bob can now be sure that with probability of $1 - \varepsilon$ the strings are equal. Because they use an ε-ASU_2 class of hash functions the collision probability for two distinct values is at most ε as the Definition 2.5 for universal hash functions declare. Also the probability for Eve to modify one of this authentication messages is at most ε, even if she suppresses modifications during the error correction phase or make Alice and Bob believe they have different strings when, in fact, they have not.

The required key material to authenticate the reconciliation phase is the sum of all costs w_1, w_2, w_3, w_4 (see Eqs. 3.15, 3.16, 3.17, and 3.18), which results in the total error correction authentication costs

$$\begin{aligned} t_{EC} &= 4(g_{EC} + \log_2 \log_2 n) \log_2 n + 4(g_{\text{auth}} + \log_2 \log_2 g_{EC}) \log_2 g_{EC} \\ &\quad + \tilde{g}_{EC} + 4(g_{\text{auth}} + \log_2 \log_2 \tilde{g}_{EC}) \log_2 \tilde{g}_{EC}. \end{aligned} \tag{3.19}$$

3.2.5 Privacy Amplification

The last step to the secret key is the privacy amplification phase. Bennett, Brassard, Crépeau, and Maurer [6] give a good explanation what privacy amplification is:

Privacy amplification is the art of distilling highly secret shared information, perhaps for use as a cryptographic key, from a larger body of shared information that is only partially secret. Let Alice and Bob be given a random variable W, as a random n-bit string, while an eavesdropper Eve learns a correlated random variable V, providing at most $t < n$ bits of information about W, i.e., $H(W|V) \geq n - t$. The details of the distribution P_{VW} are generally unknown to Alice and Bob, except that it satisfies this constraint as well as possibly some further constraints. They may or may not know P_W. Alice and Bob wish to publicly choose a compression function $g : \{0, 1\}^n \rightarrow \{0, 1\}^r$ such that Eve's partial information on W and her complete information on g gives her arbitrarily little information about $K = g(W)$, except with negligible probability (over possible choices for g). The resulting K is virtually uniformly distributed, given all Eve's information; it can hence be used safely as a cryptographic key.

The size r of the secret that Alice and Bob can distill depends on the kind as well as the amount of information available to Eve. Eve can obtain

- t arbitrary bits of W,
- t arbitrary parity checks of W,
- the result of an arbitrary function mapping n-bit strings to t-bit strings,
- the string W transmitted through a binary symmetric channel with bit error probability p satisfying $h(p) = 1 - \frac{t}{n}$, and hence with capacity $\frac{t}{n}$ where h denotes the binary entropy function.

Assume that we have the scenario where Alice and Bob are connected by an insecure channel to which Eve has passive perfect access. Note that with authentication of the communication active perfect access has the same characteristics, which will not be discussed here for reasons of simplicity. So after Alice's transmission on the quantum channel each of them has knowledge of a correlated random variable. Alice X, Bob Y, and Eve Z. Assume that these variables are distributed according to some joint probability function p_{XYZ}, whereby Eve has partially control over this distribution. Based on the fact that neither Alice nor Bob has an advantage compared to Eve concerning information ($I(X;Y) \not> I(X;Z)$, $I(X;Y) \not> I(Y;Z)$), after the sifting phase which can be denoted as random variable C, Alice can compute a string W from X and C such that Alice's uncertainty about W is 0 and Bob's uncertainty about W is smaller than Eve's one:

$$\begin{aligned} H(W|XC) &= 0, \\ H(W|YC) &< H(W|ZC). \end{aligned} \tag{3.20}$$

The reconciliation phase helps to remove Bob's uncertainty about W; therefore, Alice sends a bit string D with length l is slightly larger than $H(W|YC)$ such that $H(W|YCD) \approx 0$. Eve's uncertainty $H(W|ZCD)$ is lower bounded by $H(W|ZC) - l$, which can be substantially positive, due to Eq. 3.20. We summarize Eve's knowledge ZCD about W to the random variable V.

The aim of privacy amplification is now to generate a secret key K, of which Eve has negligible amount of information. Therefore, Alice and Bob agree on a function g (known by Eve) to generate $K = g(W)$. This g is chosen randomly from a set $\mathcal{G}$ of function to avoid that Eve can decide beforehand which strategy to use. Thus, this function is also a random variable G on the set $\mathcal{G}$. The output length of $G\colon W \rightarrow \{0, 1\}^r$ should decrease Eve's information about K to a minimum. Assume $I(W;V) \leq t$ then $I(K;GV) \approx 0$ and $H(K|GV) \approx r$. But how must r be chosen? In

[6] they state $r = n - t - s$ and the security parameter s decreases Eve's information rapidly, because $I(K;GV) \leq \frac{2^{-s}}{\ln 2}$.

The theorem and the corollaries of [6] are presented now, which yields an upper bound to Eve's information. It emerges that for the function set $\mathcal{G}$ the universal classes of hash functions as described in Sect. 2.2.1 become useful, due to their properties.

Theorem 3.2 *Let X be a random variable over the alphabet $\mathcal{X}$ with probability distribution p_X and Rényi entropy $R(X)$, let $Q = G(X)$. Then*

$$H(Q|G) \geq R(Q|G) \geq r - \log_2\left(1 + 2^{r-R(X)}\right) \geq r - \frac{2^{r-R(X)}}{\ln 2},$$

where G is the random variable corresponding to the random choice (with uniform distribution) of a member of a universal class of hash functions $\mathcal{X} \to \{0, 1\}^r$.

The first inequality follows from Eq. 2.45. The second inequality can be proven by using Jensen's inequality and using the fact that we use universal hash function. For the complete proof we refer to [6]. The last inequality follows from $\log_2(1+y) \leq \frac{y}{\ln 2}$. The next corollary is derived from the above theorem.

Corollary 3.1 *Let P_{VW} be an arbitrary probability distribution and let v be a particular value of V observed by Eve. If Eve's Rényi entropy $R(W|V = v)$ about W is known to be at least c and Alice and Bob choose $K = G(W)$ as their secret key, then*

$$H(K|G, V = v) \geq r - \log_2\left(1 + 2^{r-c}\right) \geq r - \frac{2^{r-c}}{\ln 2},$$

where G is chosen at random from a universal class of hash functions $W \to \{0, 1\}^r$.

Corollary 3.2 *Let W be a random n-bit string with uniform distribution over $\{0, 1\}^n$, let $V = e(W)$ for an arbitrary eavesdropping function $e : \{0, 1\}^n \to \{0, 1\}^t$ for some $t < n$, let $s < n - t$ be a positive safety parameter, and let $r = n - t - s$. If Alice and Bob choose $K = G(W)$ as their secret key, then Eve's expected information about the secret key K, given G and V, satisfies*

$$I(K;GV) \leq \frac{2^{-s}}{\ln 2},$$

where G is chosen at random from a universal class of hash functions $\{0, 1\}^n \to \{0, 1\}^r$.

Finally, corollary 3.2 can be proven with usage of Corollary 3.1 (see [6]). Note that the security parameter does not depend on the error probability p or else variables, if the upper bound for Eve's knowledge t is known.

Let us recapture the entire protocol to identify the knowledge t Eve has obtained. The easiest knowledge to identify is in the reconciliation phase and the authentication of the reconciliation. We know that during error correction at least $nh(p)$ bits (p is the error probability) are necessary to correct all errors. In Sect. 3.2.3.2 a reconciliation protocol has been presented which reaches nearly this limit. It differs only by a factor $f(p)$ given as in Table 3.3. Thus, the knowledge Eve accumulates during this phase is $nf(p)h(p)$. If we formulate this as a fraction τ_1 by which the reconciled key is to shorten then

$$\tau_1(p) = \frac{nf(p)h(p)}{n} = f(p)h(p) = -f(p)\left(p\log_2 p + (1-p)\log_2(1-p)\right). \tag{3.21}$$

In the confirmation/authentication of the error correction phase a hash value of the key is transmitted. This tag has length g_{EC}, giving Eve additional information. The last phase where Eve can gain knowledge is the sifting phase, if she eavesdrops several photons during quantum transmission. How much information Eve gets, depends on the error probability p. Lütkenhaus derived in [14] and [15] a fraction τ_2, by which we need to shorten the reconciliation key in addition to the other parameters to get a secure key. Therefore, he used a trivial extension of Corollary 3.1:

$$I(K;GV) = H(K) - H(K|GV) = r - H(K|GV) \leq r - R(K|GV). \tag{3.22}$$

This inequality can be applied due to Eq. 2.45. Furthermore, in [14] it is shown that the collision probability of a key k is bounded above by the collision probability of his sifted key for averaged g and v, hence

$$I(K;GV) \leq r - R(K|GV) \leq r - R(W|V). \tag{3.23}$$

We set the information $I(K;GV)$ to zero and thus the shortening is

$$\tau_2 = \frac{n-r}{n} = 1 - \frac{R(W|V)}{n} = 1 + \log_2\langle p_c^w(v)\rangle_v, \tag{3.24}$$

where $\langle\rangle_v$ means averaged with respect to v and $p_c^w(v)$ is the collision probability $\sum_{w\in\mathcal{W}} p(w|y)^2$. We can express the collision probability for the sifted key as the product of the collision probability for single bits of this sifted key $\langle p_c^w(v)\rangle_v = (p_c^{(1)}(p))^n$ with respect to the error rate of the quantum channel p. In [14] this collision probability was derived as

$$p_c^{(1)}(p) \leq \frac{1}{2} + 2p - 2p^2 \qquad \text{for } p \leq 1/2, \tag{3.25}$$

which gives, finally,

$$\tau_2(p) \leq \log_2\left(1 + 4p - 4p^2\right) \qquad \text{for } p \leq 1/2. \tag{3.26}$$

The gain for the privacy amplification g_{pa} is composed of τ_1 from Eq. 3.21, τ_2 from Eq. 3.26, $\frac{s}{n}$ (security parameter length) and $\frac{g_{EC}}{n}$ (authentication tag length)

$$\begin{aligned} g_{pa} &= 1 - \tau_1(p) - \tau_2(p) - \frac{s}{n} - \frac{g_{EC}}{n} \\ &= 1 + f(p)(p\log_2 p + (1-p)\log_2(1-p)) \\ &\quad - \log_2\left(1 + 4p - 4p^2\right) - \frac{s}{n} - \frac{g_{EC}}{n} \qquad \text{for } p \leq 1/2. \end{aligned} \tag{3.27}$$

Note that the two fractions $\frac{s}{n}$ and $\frac{g_{EC}}{n}$ are not crucial factors for the final key, because they are almost fixed values and if the length of the key increases they are comparatively small. Figure 3.7 shows that the maximal acceptable error rate is around 10% for CASCADE and around 11% if we reach Shannon's limit. The factors $\frac{s}{n}$ and $\frac{g_{EC}}{n}$ are not considered.

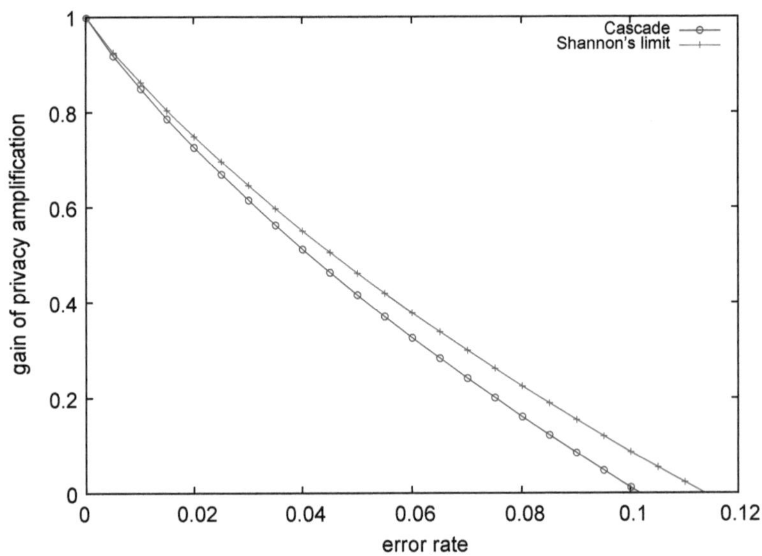

Fig. 3.7 The gain of privacy amplification g_{pa} for increasing quantum error rate p

Concluding the QKD protocol we give a universal class of hash function, which can be used to reduce the reconciliation key of length n to a final key of length k. Therefore, the multiplication in finite fields is defined as

Definition 3.6 Let $\mathcal{A} = GF(2^n)$ and $\mathcal{B} = \{0, 1\}^k$. Let $h_c(x)$ be defined as the last k bits of the product $c \cdot x$ in a polynomial representation of $GF(2^n)$. The set

$$\mathcal{H}_{GF(2^n)\rightarrow\{0,1\}^k} = \{h_c : c \in GF(2^n)\}$$

is a universal family of hash functions [23].

This family requires only n bits to identify the particular function (the value of c). In particular it is chosen for privacy amplification because the multiplication of large blocks can be done efficiently. We know that by using the traditional shift-and-add algorithm, the multiplication of two elements of $GF(2^n)$ can be achieved in quadratic time. In [2] a simple algorithm is presented, which performs a multiplication in $a \cdot l \log l$ steps for some small constant a.

The value c, which chooses a function of the family of hash functions, must be random. Alice or Bob could randomly choose such a value c and announces it publictly to its peer. But we must consider that this data exchange must be authenticated, not to be exposed to an attack by Eve. Recalling the sifting phase, where Bob has sent Alice an authenticated message, in which he told her, which photons he was able to measure and what bases he has used. A convention was that Bob chooses his bases randomly, so Alice and Bob share a $2n$ authenticated random string (the length is $2n$ because normal sifting has a gain of 0.5). They can now extract the first n bits of this string and use it as c. For this reason the communication during the privacy amplification phase and thus an additional authentication is not necessary.

As an example let the string for the random choice of base, which Bob sent authenticated to Alice, be 00100111010110001 01. After performing sifting and reconciliation, the reconciliation key, shared by Alice and Bob, is 1101001110. Due to the error rate and the information leaked during reconciliation we have computed that Eve has 4 bits of information, so we must shorten the key to the length of 6 bits. We extract the value of c from Bob's random choice of bases, which is 0010011101 and multiply c with the reconciliation key.

$$0010011101 \cdot 1101001110 = 1000000110110 10110$$

The final key consists now of the last six bits of the result (010110).

3.3 QKD Gain

In Sect. 3.1 and 3.2 discussing the communication over the classical and the quantum channel we heard how a QKD system works and some equations are given to compute the output of the several stages. The gain of the quantum channel g_q in Eq. (3.1), the protocol gain g_p in Eq. (3.2), and the gain of privacy amplification g_{pa} in Eq. (3.27) form the complete gain. Additionally we must subtract the required authentication key (t_S (3.6) and t_{EC} (3.19)) which are needed for the next round. So the length k of the final authenticated key for one round is

$$k = m \cdot g_q \cdot g_p \cdot g_{pa} - t_S - t_{EC}, \tag{3.28}$$

where m is the length of the random sequence Alice generates at the beginning and sends via the quantum channel to Bob. The major problem with this protocol is that for the first round we need a pre-shared secret between Alice and Bob. Thus, the key output length would be limited to the length of the secret. After a certain "warm-up" phase the protocol would reach the expected output on average.

3.4 Finite Resources

All assumptions regarding the postprocessing (sifting, error estimation and correction, privacy amplification) and hence the security of quantum key distribution protocols have been proven in the asymptotic limit. However, the actual implementations of the protocols can only use finite resources, such as limited computational power and keys of finite length. This fact has already been addressed for special protocols like the BB84 protocol [12] or the six-state protocol [16]. In both scenarios it has been shown that the values for the sifted key rate in the finite scenario differ significantly from the expected values from the asymptotic limit.

As described in [20] also the definition of security has to be altered when considering finite resources. In the asymptotic limit a key of length l is said to be secure if its deviation ε from a perfect key tends to zero as l increases. Another property most security definitions lack is composability. Composability assures that a key coming from a quantum cryptographic protocol can safely be used in classical applications (e.g., for encryption with a one-time pad). For the nonasymptotic scenario Scarani and Renner presented a security definition, which satisfies also composability [20]. Here a key is considered to be secure if the difference between the generated key and a perfect key is smaller than ε, which means that ε is the maximum probability that the generated key differs from a perfect key.

The main goal Scarani and Renner pursue in their paper is to obtain a sifted key rate r based on a certain number of signals, a security parameter ε, and some losses from the error correction. As starting point they take the relation from the asymptotic limit, where the sifted key rate r' is

$$r' = S(X|E) - H(X|Y), \tag{3.29}$$

where S is the von Neumann and H the Shannon entropy. The goal is achieved using tools from Renner's PhD thesis [18]. The exact formulas and proofs are described in [20] and will not be further discussed here. The major result is that the key rate becomes positive, e.g., for a little more than 10^4 signals and an error rate of 0.5% or at least 10^5 signals and an error rate of 5%.

References

1. Ardehali, M., Chau, H.F., Lo, H.K.: Efficient quantum key distribution (1998). URL http://www.citebase.org/abstract?id=oai:arXiv.org:quant-ph/98% 03007
2. Assche, G.V.: Quantum Cryptography and Secret-Key Distillation. Cambridge University Press, New York, USA (2006)

3. Bennett, C.H.: Quantum cryptography using any two nonorthogonal states. Phys. Rev. Lett. **68**(21), 3121–3124 (1992). DOI 10.1103/PhysRevLett.68.3121
4. Bennett, C.H., Bessette, F., Brassard, G., Salvail, L., Smolin, J.A.: Experimental quantum cryptography. J. Cryptology **5**(1), 3–28 (1992)
5. Bennett, C.H., Brassard, G.: Quantum cryptography : Public key distribution and coin tossing. In: Proceedings of the IEEE International Conference on Computers, Systems, and Signal Processing, pp. 175–179 (1984)
6. Bennett, C.H., Brassard, G., Crépeau, C., Maurer, U.M.: Generalized privacy amplification. IEEE Trans. Inf. Theory **41**(6), 1915–1923 (1995)
7. Brassard, G., Salvail, L.: Secret-key reconciliation by public discussion. In: EUROCRYPT, pp. 410–423 (1993)
8. Bruss, D.: Optimal eavesdropping in quantum cryptography with six states. Phys. Rev. Lett **81**(14), 3018–3021 (1998)
9. Carter, L., Wegman, M.N.: Universal classes of hash functions. J. Comput. Syst. Sci. **18**(2), 143–154 (1979)
10. Ekert, A.K.: Quantum cryptography based on bell's theorem. Phys. Rev. Lett. **67**(6), 661–663 (1991). DOI 10.1103/PhysRevLett.67.661
11. Gilbert, G., Hamrick, M.: Practical quantum cryptography: A comprehensive analysis (part one) (2000). URL http://www.citebase.org/abstract?id=oai:arXiv.org:quant-ph/00% 09027
12. Inamori, H., Lütkenhaus, N., Mayers, D.: Unconditional security of practical quantum key distribution. Eur. Phys. J. D **41**(3), 599–627 (2007)
13. Lo, H.K., Chau, H.F., Ardehali, M.: Efficient quantum key distribution scheme and proof of its unconditional security. Journal of Cryptology **18**, 133 (2005). URL http://www.citebase.org/abstract?id=oai:arXiv.org:quant-ph/0011056
14. Lütkenhaus, N.: Estimates for practical quantum cryptography. Phys. Rev. A **59**, 3301 (1999). URL http://www.citebase.org/abstract?id=oai:arXiv.org:quant-ph/9806008
15. Lütkenhaus, N.: Security against individual attacks for realistic quantum key distribution. Phys. Rev. A **61**(5), 052,304 (2000). DOI 10.1103/PhysRevA.61.052304
16. Meyer, T., Kampermann, H., Kleinmann, M., Bru, D.: Finite key analysis for symmetric attacks in quantum key distribution. Phys. Rev. A **74**(4), 042,340 (2006)
17. Nielsen, M.A., Chuang, I.L.: Quantum Computation and Quantum Information. Cambridge University Press, Cambridge (2000). URL http://www.amazon.ca/exec/obidos/redirect?tag=citeulike09-20% &path=ASIN/0521635039
18. Renner, R.: Security of Quantum Key Distribution. Ph.D. thesis, Swiss Federal Institute of Technology
19. Scarani, V., Acin, A., Ribordy, G., Gisin, N.: Quantum cryptography protocols robust against photon number splitting attacks for weak laser pulses implementations. Phy. Rev. Lett. **92**(5), 057,901 (2004)
20. Scarani, V., Renner, R.: Quantum cryptography with finite resources: Unconditional security bound for discrete-variable protocols with one-way postprocessing. Phys. Rev. Lett. **100**(20), 200,501 (2008)
21. Smith, G., Renes, J.M., Smolin, J.A.: Better codes for BB84 with one-way post-processing (2006). URL http://www.citebase.org/abstract?id=oai:arXiv.org:quant-ph/0607018
22. Tang, X., Ma, L., Mink, A., Nakassis, A., Xu, H., Hershmanand J. Bienfang, B., Su, D., Boisvert, R.F., Clark, C., Williams, C.: Quantum key distribution system operating at sifted-key rate over 4 Mbit/s. In: Quantum Information and Computation IV., *Presented at the Society of Photo-Optical Instrumentation Engineers (SPIE) Conference*, Vol. 6244 (2006). DOI 10.1117/12.664455
23. Wegman, M.N., Carter, L.: New hash functions and their use in authentication and set equality. J. Comput. Syst. Sci. **22**(3), 265–279 (1981)
24. Xu, H., Ma, L., Mink, A., Hershman, B., Tang, X.: 1310-nm quantum key distribution system with up-conversion pump wavelength at 1550 nm. Optics Express **15**, 7247–7260 (2007)

Chapter 4
Adaptive Cascade

S. Rass and C. Kollmitzer

4.1 Introduction

Quantum cryptographic key exchange is a promising technology for future secret transmission, which avoids computational infeasibility assumptions, while (almost) not presuming pre-shared secrets to be available in each peer's machine. Nevertheless, a modest amount of pre-shared secret information is required in adjacent link devices, but this information is only needed for authentication purposes. So quantum key distribution cannot create keys from nothing, rather it is a method of key expansion. The remarkable feature of quantum cryptography is its ability to detect eavesdropping by the incident of an unnaturally high quantum bit error rate. On the other hand, it has no defense against person-in-the-middle attacks by itself, which is why authentication is of crucial importance.

We are particularly interested in the final steps of the BB84 protocol, in which Alice and Bob correct errors in the bit strings they created from their measurements. Key reconciliation involves Alice and Bob publicly exchanging parity bits in order to correct errors and distill identical keys.

4.2 Error Correction and the Cascade Protocol

Coming to the core topic of this chapter, let us pay closer attention to the error correction mechanism which has been proposed along with the experimental implementation of BB84 [2]. Errors in physical transmission media often exhibit burst structures, that is, a sequence of consecutive errors is more likely to occur than

S. Rass (✉)
System Security Research Group, Institute of Applied Informatics, Universitaet Klagenfurt, Universitaetsstrasse 65-67, 9020 Klagenfurt, Austria, stefan.rass@uni-klu.ac.at

C. Kollmitzer
Safety & Security Department, Quantum Technologies, AIT Austrian Institute of Technology GmbH, Lakeside B01A 9020 Klagenfurt, Austria, christian.kollmitzer@ait.ac.at; http://www.ait.ac.at

Rass, S., Kollmitzer, C.: *Adaptive Cascade*. Lect. Notes Phys. **797**, 49–69 (2010)
DOI 10.1007/978-3-642-04831-9_4 © Springer-Verlag Berlin Heidelberg 2010

sparse scattering. Consequently, a popular trick is to permute the bits prior to any error correction in order to chop down long bursts into small pieces, ideally leaving an almost uniform pattern of erroneous bits in the result. This is the first step in the Cascade protocol that has been summarized in Chap. 3. After having agreed on a publicly known permutation of bits, Alice and Bob take their shuffled strings and partition it into blocks of size k, such that a single block is believed to contain no more than one error with high probability. In [2], a theoretical treatment of the optimal choice of block size is missing, and this is precisely the gap we intend to close here. The problem of how errors are scattered across the raw key has been tackled on statistical grounds before [17]. For this analysis, a binomial distribution of errors within the blocks is assumed, which is later approximated by a Poisson distribution. We shall take a different route here, considering the process that induces the errors to be Poissonian, as well as adapting the initial block-size using a decision-theoretic approach. Moreover, the authors of [17] do not provide a direct clue on how to choose an optimal block-size. Before getting into details about how to cleverly choose the block-sizes, let us outline the remaining steps in the error correction process, as this will illustrate the efficiency gain we may expect from an intelligent partitioning strategy. The details and some theoretical results about the process have been summarized in Chap. 3, but we briefly repeat them here for convenience of the reader.

Having split the string into blocks of equal size k, Alice and Bob publicly compare parity bits of each block. Obviously, one error will change the parity, and in general, any odd number of errors will be discovered by observing disagreeing parities. However, two or any larger even number of errors will remain undetected with this method, which is why further stages of the process are to follow, once the initial correction has been completed. Let us describe the general correction step by an example block with one indicated error, i.e., unequal parity by public comparison. Then this block is searched for the error using a standard bisective search, which discloses a further lot of $\log(k)$ parities of sub-blocks. To spot and correct remaining errors in the string, such as present in blocks with an even number of errors in them, Alice and Bob repeat the randomization and partitioning steps, several times with increasing block-sizes.

Since the error correction up to now may be ineffective, as still having missed some errors, Alice and Bob continue by comparing parities of random subsets of bits they publicly agree on. Upon parity mismatch, a bisective search similarly as above is performed to find and erase the error. In order to avoid information leaking to the adversary, the last bit from each random subset can be discarded. This deletion can also be done after comparing parities of blocks in the previous steps for the same reason.

The point at which almost all errors have been removed is detected by counting the number of successful comparisons after having seen the last error. After a sufficient number of successful trials (20 is the number proposed in [2]), the strings are accepted as identical, regarding the probability of errors remaining undetected as negligible.

The protocol Cascade is based on this procedure and has been introduced in a later paper [5], which presented improvements to the original reconciliation protocol sketched above. Among the changes is the replacement of the bit-deletion step by a smarter strategy for the sake of detecting more errors faster, so the task of information leakage reduction is shifted to the privacy amplification stage. The naming stems from the strategy of increasing sizes of blocks in the first stages of the protocol. Although a comprehensive theoretical analysis of the protocol is provided, the authors of [5], as well as of [2], abstain from an analytical treatment of block-size choices. Nevertheless, a simple heuristic based on estimating the error frequency by direct comparison of a random sample of bits is provided in [2]. These bits have to be sacrificed for the sake of privacy too, if that approach is adopted. Our main concern in subsequent sections will thus be an optimal choice of block-sizes in the initial stage of the protocol. Since the increase of block-sizes may be taken exponential (taking double sizes in stage two, quarter-size blocks in stage three of Cascade, and so on), the optimal choice of block-size in terms of deriving deterministic or probabilistic models may save us from too many stages of Cascade after the initial one. This concern has been pointed out in [2] already, since the block parity comparison approach becomes much less efficient the larger the blocks grow. The idea of estimating the error frequency from a random sample is interesting though, as this is already some kind of adaption to the current situation rather than taking a constant block-size once and for all. In the following, we show how this idea can be amended to spare the need for sacrificing too much key material.

4.2.1 Sources of Errors

It is worthwhile to lose a few words about natural sources of errors (considering an adversary as a rather unnatural source of such phenomena), since this is the type of errors we need to model for an improved estimate of optimal block-sizes. A simple systematic error pattern arises from photon emitters and detectors slowly running out of synchrony. It is easy to imagine that if a photon emitter sends photons with frequency f and the corresponding detector samples with frequency $f+\varepsilon$ for some small $\varepsilon > 0$, then we will observe an oscillatory pattern of error bursts, i.e., we do not sample anything at all over repeating periods of time. Furthermore, environmental conditions (heat, vibration, etc.) may have an impact on the transmission medium and hence influence the error rate. Since photon emitters and detectors are devices sensitive to calibration errors, increased vibrations due to traffic peak hours may distort the setup and hence increase the error rate. The form of such perturbations may be modeled well, and is perhaps even periodic to some extent, hence can be captured by partially deterministic models as described below.

Further sources of errors may be related to the quantum transmission medium. Sending photons over fiber-optic lines, we may observe error frequencies that are essentially different from those seen on free-space links. The latter are subject to various environmental distortions, including weather conditions (fog, rain, snow, etc.), air pollution, or radiation. Kinks or points where fiber-optic cables are joined by splicing or cleaving may increase photon loss and hence the error rate. Damages caused during laying the cables are sometimes unavoidable and need to be taken into account as another source of error. Common to both variants is its damping that is dependent on the length of the channel, which amounts to less photons making it through, the longer the line gets. A comprehensive overview of system losses is provided in [17].

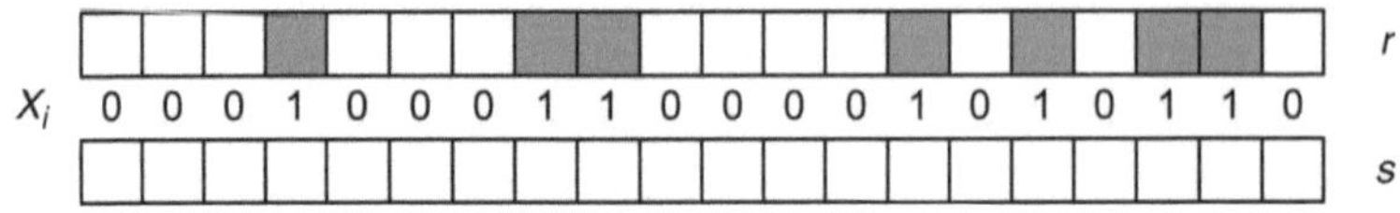

Fig. 4.1 Example distribution of random errors over the raw key r. Bits marked in *gray* are erroneous and thus not carried over to the sifted key s

4.3 Adaptive Initial Block-Size Selection

The original Cascade protocol, based on the ideas sketched above, offers room for improvement in terms of its initial block-size selection strategy. We shall not deal with adaptive choices of block-sizes in later stages of Cascade than the first one, so the term *block-size* henceforth refers to the initial block-size, unless stated otherwise. In [2], the choice of block-size is based on a prior estimate of error frequency so as to achieve an expected number of one error per block; why not incorporate the posterior knowledge from a finished protocol run in future executions? This appears reasonable, since we can assume the link to be geographically static, so undergoing the same sort of distortions regularly. We hence may be able to continuously improve our error estimate over the lifetime of the link devices. Let us first summarize what information is available after one execution of the protocol: call $r \in \{0, 1\}^*$ the raw key being the input for the Cascade protocol. After completion of the error correction, and before the privacy amplification, we have a string s of the same length as r, but with errors being corrected. Checking for each bit in r whether or not it has made it into the corrected string s *without a change* gives rise to a sample from a binary-valued random variable or, more specifically, a binary-valued stochastic chain, where the discrete time refers to the index of the bits in r. Since r is itself finite, so is the time horizon of the stochastic chain, and for simplicity, we may consider it a simple random variable in the first instance. Call this random variable $X \in \{0, 1\}$, then we observe

$$X_i = \begin{cases} 1, & \text{if the } i\text{th bit in } r \text{ became corrected and has thus changed in } s, \\ 0, & \text{otherwise.} \end{cases}$$

Figure 4.1 shows an example of the distribution of errors over a bit string.

Our goal is replacing the static error frequency estimate at the beginning of the Cascade protocol by a dynamic one, which takes into account information from previous protocol executions. A variety of different solutions can be imagined, and we shall summarize a selection here, restricting ourselves to two high-level approaches:

1. Choose the block-size at the beginning and leave it constant over the whole execution time of Cascade's first stage, resulting in an equidistant partitioning. The initial choice is made using past knowledge about error distributions. We refer to this approach as *with fixed initial block-size*. This variant is suitable if the error rate remains constant over short periods, but varies in the long run (due to slow seasonal variations). The resulting scheme will hence be the choice either for quick establishment of short keys or in situations where conditions can artificially be kept almost constant over a long time. We describe two sub-variants of how to incorporate past experience in order to reach an improved idea of how much error to expect in future executions.
2. If the error rates are frequently varying within the duration of the protocol execution, then we may adapt it in real time during the first stage. The resulting model can either be a partially deterministic one, if the error rates exhibit repeating patterns (i.e., some kind of periodicity), or completely stochastic, if no such underlying pattern can be identified. In case we are dealing with fully non-deterministic scattering of errors, we propose using Bayesian statistics for adapting the error-rate model to the particular environmental conditions of a link. In that sense, we can endow Cascade with *learning capabilities* in order to self-adapt to changing conditions. We refer to this variant as *with dynamic initial block-size*.

Among our main concerns will be efficiency, since quantum cryptographic key exchange protocols can be expected to run endlessly, continuously establishing key material for ad hoc usage. Efficiency of the protocols is hence a major issue. Let us start with the simpler of the methods outlined above, choosing its block-size constant over the whole execution time. Fine-tuning of any of the dynamic models described here can be done by simulations, comparing the generated scattering to the true error patterns. It is thus useful to describe how error patterns can be simulated using the dynamic variants. We do this below.

4.4 Fixed Initial Block-Size

Treat the vector of realizations of X_i as a sample and call λ its arithmetic mean. Moreover, assume that from n previous executions of the protocol, we have the data $\lambda_1, \ldots, \lambda_n$ available. Each λ_i represents the average number of errors in the ith execution of the protocol. This average number is called the *error rate*. Furthermore, this is also the maximum likelihood estimator for the parameter of the Poisson and exponential distribution, which will become useful later. The set of (consecutive) λ_i can be considered as a time series, and our wish is to predict its next value, to use it as our error estimate for running the Cascade protocol. A considerable amount of literature exists on smoothing and trend estimation in time series, and we refer the interested reader to [6, 40] for an introduction and overview. Here, we shall only mention three popular variants, leaving the reader the freedom to apply more sophisticated methods.

To reduce the random distortions on a time series (smoothing), we may for instance

1. use sliding centrality measures, like the sliding mean or median, or
2. use exponential smoothing, taking into account all past measurements, or
3. if the state (in terms of error frequency) of the quantum channel can be approximated by a (linear) dynamical system, then the Kalman filter may be used as a predictor for the error rates in future executions. Similarly, the particle filter is an option, if a stochastic model underneath the error distribution can be set up effectively. However, both options strongly rely on the accuracy of the model and presume thorough analysis and knowledge of the system dynamics. Furthermore, the particle filter may come at high computational cost if it is to be implemented effectively.

Applying other techniques is possible, but beyond the scope of this chapter. We leave this to the reader, consulting the cited references. Coming back to the first among the above three proposals, the arithmetic mean appears to be not the optimal choice, as being very sensitive to outliers. Outliers in our setup would be short-term deviations of the error frequency by conditions becoming almost ideal for a short while or by suffering from strong environmental influences like vibrations, heat, or similar for a limited period of time. Both events may happen on a regular basis and should hence not be considered as outliers. However, the arithmetic mean estimate would strongly react on this and the resulting block-size could be smaller than needed. Detection and elimination of outliers is a highly nontrivial task, and no panacea solution is likely to exist.

A sliding median is more robust against outliers and offers a second appealing feature, as the window size has a particularly simple interpretation if the median is used: suppose that our window is of size $n = 2k + 1$ for an integer $k \geq 1$, then we consider an extreme event to be an outlier, unless it was observed at least $k + 1$ times over the last $2k + 1$ executions of the protocol. For example, if $n = 21$, then any extreme situation is considered exceptional (an outlier), unless it occurred at least 11 times during the last 21 executions. If an expert can somehow estimate the likelihood of such extremal conditions (by using extreme value probability models), then the window size can be chosen to handle that probability and hence become stable against outliers. Implementing more complex and powerful models like ARMA (autoregressive moving average) is possible, but the quality of results is highly dependent on the expertise of the engineer, as the modeling becomes involved for these techniques. See [31] for an introduction and details.

If a sliding centrality measure is employed, then the resulting formulas for choosing the block-size at the beginning are simple, as we only have to take the median (or mean or any other centrality measure of choice) over the last n available samples X_i.

Exponential smoothing [6] does include all past observations. The predicted value $\hat{\lambda}_{n+1}$ is found as $\hat{\lambda}_{n+1} = \mu\lambda_n + (1-\mu)\hat{\lambda}_{n-1}$, where $0 < \mu < 1$ is the smoothing parameter and $\hat{\lambda}_1, \ldots, \hat{\lambda}_{n-1}$ are the so-far smoothed values, setting the first $\hat{\lambda}_1 = \lambda_1$. The reader will instantly verify that this corresponds to a weighted sum of observations with exponentially decaying weights.

A further alternative is offered by decision theory: call the error rate λ and assume it to be random, then our problem is to find the best decision on what λ to choose next before re-running the protocol. Why not perform a Bayesian estimation of it? A fully natural approach to updating λ is considering it to be drawn from a random variable Λ, whose distribution is to be estimated upon our given information. Again, let the information available be the average error rate known from the last protocol execution. Furthermore, for the sake of analytical simplicity and efficiency, let us model the random variable Λ using a Gamma distribution with hyper-parameters $a, b > 0$, which are to be specified by an expert prior to any protocol runs.

Recall that we update the distribution of Λ after each protocol execution, which provides us with only a single sample, being the average error rate. Assuming that this rate is Poisson-distributed, the likelihood function for a single sample is trivially a Poisson density. The reader familiar with Bayesian decision theory will instantly recognize the Gamma distribution as a conjugate prior for a Poissonian likelihood function, in which case a Bayesian update can efficiently be computed [37]. To make this precise, updating the Gamma distribution

$$\mathcal{G}(a,b) \quad \text{with density} \quad f(\lambda|a,b) = \begin{cases} \frac{b^a}{\Gamma(a)}\lambda^{a-1}e^{-b\lambda}, & \text{if } \lambda \geq 0 \\ 0, & \text{if } \lambda < 0, \end{cases}$$

to account for the information z (being the average number of errors derived from the most recent measurements) gives the posterior distribution

$$\mathcal{G}(a+z, b+1), \tag{4.1}$$

so the update amounts to a simple change of the parameters. Using a quadratic loss function, i.e., penalizing a deviation of the estimator $\hat{\lambda}$ from its true value λ by the functional $(\lambda - \hat{\lambda})^2$, the Bayesian estimator $\hat{\lambda}$ minimizing the loss we can expect under the posterior estimate of the parameter (given by (4.1)) is just the expectation of the (posterior) Gamma distribution and found as

$$\hat{\lambda} = \mathbb{E}[\Lambda|z] = \frac{a+z}{b+1},$$

referring to the parameters of the posterior distribution. A popular argument against conjugate priors is their inflexibility. A straightforward remedy is using a mixture of conjugate priors for a close approximation of a more satisfactory prior, but although the set of mixtures of priors can be shown to be dense in the set of probability distributions (with respect to the Prohorov topology) [37, Theorem 3.4.3], its usage introduces the need for normalization after the Bayesian update. This means that we are back at the lengthy calculations involved when using Monte Carlo methods (see [37, 15] for an introduction) for evaluating the integrals occurring in the expressions, and hence this option may be taken if one is willing to trade efficiency for accuracy.

4.5 Dynamic Initial Block-Size

If changes in the error frequency can exhibit regular (perhaps periodic) behavior, then we may approximate these rates using deterministic models that we learn from the available information. Consequently, we may set up a stochastic Poisson process [19, 24] with deterministic intensity function (i.e., an inhomogeneous process) or use a fully non-deterministic model for the rates, giving a Cox process model for the error scattering. Each approach is expanded below, starting with the determination of a model for given empirical error rates.

4.5.1 Deterministic Error-Rate Models

Due to the structure of the Cascade error correction method, we have blocks with errors located inside, and the number of these is available after the protocol terminates. Enumerate the blocks by $i = 1, 2, \ldots, N$ and call t_i the time stamp of the transmission where the middle bit in the ith block has been sent. Let λ_i denote the number of errors within the ith block divided by the length (time span) of the ith block, then we are left with a classical curve-fitting problem upon a given data set (t_i, λ_i), for $i = 1, \ldots, N$. This can be solved by standard techniques. Updating an existing estimate Λ upon arrival of new measurements is computationally efficient, since the formula of Sherman and Morrison [39] applies. Let the fitted model be a function $\Lambda \in \text{span}(g_1, \ldots, g_m)$ for m linearly independent base functions $g_1, \ldots, g_m : \mathbb{R}^+ \to \mathbb{R}$. Possible choices for the functions g_i include the polynomial base $1, x, x^2, \ldots, x^{m-1}$ or the trigonometric base $1, \sin(x), \cos(x), \sin(2x), \cos(2x), \ldots$, which corresponds to an approximation similar as with a truncated Fourier series. Our goal is finding a function $\Lambda(t)$, which approximates the time-dependent variation of errors, assuming that this function is characteristic for the error distribution on a link. Figure 4.8 displays an example. Approximation of the given point set can be done with respect to any metric, but we propose using the Euclidian distance for the sake of an efficient update upon new data. The technique is better known as *least-squares fitting*. The standard least-squares approximation problem is the following: given a set of points (t_i, λ_i) for $i = 1, 2, \ldots, N$, find a vector of coefficients $\mu = (\mu_1, \ldots, \mu_m)$, which minimizes the squared distance functional

$$Q(\mu) = \sum_{i=1}^{N} [\lambda_i - \Lambda(t_i)]^2 = \sum_{i=1}^{N} \left[\lambda_i - \sum_{j=1}^{m} \mu_j g_j(t_i) \right]^2,$$

where

$$\Lambda(t) = \mu^T g = \sum_{i=1}^{m} \mu_i g_i(t), \tag{4.2}$$

with $g = (g_1, \ldots, g_m)$.

The solution μ to a least-squares fitting problem is then found by setting the first-order partial derivatives of Q with respect to $\mu_1, \ldots, \mu_m$ to zero. Solving for μ then gives the well-known expression

$$\mu = (A^T A)^{-1} A^T \lambda, \tag{4.3}$$

where the matrix A is called the *design matrix* and defined as

$$A = \begin{pmatrix} g_1(t_1) & g_2(t_1) & \cdots & g_m(t_1) \\ g_1(t_2) & g_2(t_2) & \cdots & g_m(t_2) \\ \vdots & \vdots & \ddots & \vdots \\ g_1(t_N) & g_2(t_N) & \cdots & g_m(t_N) \end{pmatrix}.$$

Incorporating new information from the $(n+1)$ th execution of the error correction protocol amounts to updating the currently estimated model $\Lambda(t)$ with the new sample $z := \lambda_{n+1}$. Let us reuse a piece of notation from statistics and write $\Lambda(t|z)$ for the *posterior model*, which incorporates the information z that we gathered from the last protocol execution. Consequently, we may call $\Lambda(t)$ the *prior model*. The corresponding design matrices differ only by one additional row referring to z, so denoting it by A_z, we have the posterior model design block matrix

$$A_z := \left(\frac{A}{z^T}\right),$$

with A being the design matrix of the prior model. For A_z, we have

$$A_z^T A_z = \left(A^T | z\right) \left(\frac{A}{z^T}\right) = A^T A + z z^T,$$

which shows that taking the new information z into account amounts to a rank-one update of $A^T A$. This particular type of modification permits the application of the Sherman–Morrison formula [39], which efficiently creates the inverse of $A_z^T A_z$ from the (known) inverse of $A^T A$. Setting $B := (A^T A)^{-1}$ gives [34]

$$(A_z^T A_z)^{-1} = B - \frac{B z z^T B}{1 + z^T B z},$$

provided that the denominator does not vanish. The posterior model $\Lambda(t|z)$ is hence found by using Eqs. (4.3) and (4.2) as

$$\Lambda(t|z) = [(A_z^T A_z)^{-1} A_z^T (\underbrace{\lambda_1, \ldots, \lambda_n}_{\substack{\text{previous}\\ \text{information}}}, \underbrace{z}_{=\lambda_{n+1}})^T]^T g.$$

The general technique applies for any model which is linear in the parameter vector, so the two choices above as well as exponential base functions may be used equally well, but may exhibit significant differences in terms of performance. An oscillatory behavior may be better approximated by a trigonometric base, for instance. It is thus up to the engineer to do the fine-tuning of the model for a specific link under consideration.

4.5.2 Going Beyond Least Squares

Among the assumptions underneath our method is *homoscedasticity* of the error rate's *errors*. Stated differently, we may think of error rates being described by a model $f + \varepsilon$ having random errors ε with constant second moment (variance), whereas f may describe an underlying trend model. This, however, may not necessarily be the case, since given a time series of error rates from the Cascade protocol, we may identify certain patterns of increased error rates, following dense error patterns, or vice versa. If error rates exhibit dependencies on their history, the assumption of homoscedasticity is violated and we may find a *heteroscedastic* model more appropriate. Figure 4.2 illustrates the difference between the two, showing a heteroscedastic model with linearly increasing variances. Specifically, an autoregressive conditional heteroscedastic (ARCH) or its generalized version, the GARCH model, may be promising candidates for a successful error model. Loosely speaking, a stochastic process (modeling the error rates) follows an ARCH model, if its expectation (conditional on its past) is zero, but its variance, being conditioned on the history of the process, nontrivially depends on the sigma field generated by the past observations [3]. This means that we explicitly model variations in the second moment of the error. A *generalized* ARCH model arises if an ARCH model is extended with dependencies on the variance of the error itself. Deterministic trends with time-dependent variances can be incorporated with this method. A well-written introduction is found in [18].

Taking a different route, we may switch from the simple models we described to a more flexible and general method, known as *generalized linear models* (see Chap. 7). Intuitively, one may think of such models as a uniform approach to several regression techniques, and the idea for our context is briefly sketched as follows: assume that the error rates (response) Λ follow a distribution that belongs to the exponential family (see [37] for a definition). This intensity Λ is written as a linear combination η of the influencing factors (predictors). The so-called *link function* g describes how the expected error rate $\mathbb{E}\Lambda$ is related to η, i.e., $\eta = g(\mathbb{E}\Lambda)$. In principle, any monotone continuous and differentiable link function can be used, but convenient choices and standard recommendations are given in the literature (see [30] for an introduction, as well as [16, 14, 28] for a more application-oriented

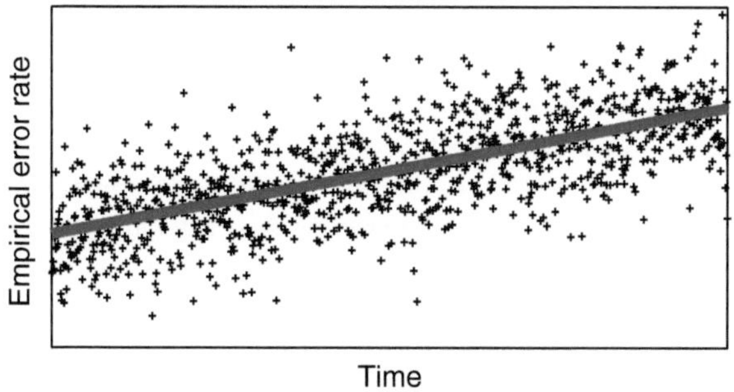

(a) Homoscedasticity

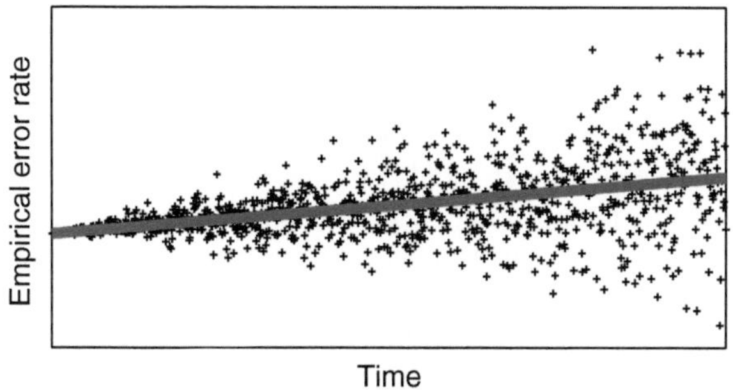

(a) Heteroscedasticity

Fig. 4.2 Example error rates with linear trend and constant/increasing measurement variances

presentation). This is particularly interesting in our case, since error rates can be classified as counting data, in which case the Poisson distribution is the natural candidate. This distribution is in the exponential family and can thus be used with generalized linear models, or *generalized linear mixed models* (GLMMs), if further flexibility is demanded.

It should be noted, however, that the mentioned extensions to the classical least-squares method often do not permit closed form analytical solutions, in which case we are stuck with intricate numerics, potentially cutting down the efficiency of the methods.

4.5.3 Error Rates as a Stochastic Process

Assuming errors to happen *independently* from each other at constant intensity λ, we may think of the event of an error as a random variable following a Poisson distribution with parameter λ. The temporal spread of errors over the raw key is then a process with independent increments also following a Poisson distribution with intensity parameter proportional to the length of the block. To make this precise, let us briefly review the properties of a Poisson process for convenience of the reader. A *stochastic process* is a family $\{X(t) : t \in T\}$ of random variables, indexed by a variable t. A standard convention interprets t as the *time*, in which case $T = \mathbb{R}^+$, i.e., $t \geq 0$. Depending on whether the variable t is continuous or discrete, we distinguish stochastic *processes* from stochastic *chains*, respectively. Moreover, the process is called *discrete* if each $X(t)$ is discrete, for otherwise it is called a *continuous stochastic process/chain* or simply a *stochastic process/chain* for short. Following the presentation of [38], a *Poisson process* is a family of discrete counting measures $\{N(t) : t \geq 0\}$, which satisfy the following three conditions:

1. $N(0) = 0$.
2. The process has independent increments.
3. The number of events in any interval of length t is Poisson-distributed with mean λt. That is, for all $s, t \geq 0$,

$$\Pr\{N(t+s) - N(t) = n\} = \frac{(\lambda t)^n}{n!} e^{-\lambda t}, \quad n = 0, 1, \ldots \quad .$$

Our Poisson process model $N(t)$ will count the total number of errors in the bit string at time t. Since our blocks should be chosen such that the expected additional number of errors after taking a time-step Δt is only 1, we are particularly interested in the distribution of inter-arrival times between two errors. It is well known that the distribution of the time between two errors is exponential with mean λ^{-1} if that parameter is constant [38]. However, this is an advantage and may be a drawback at the same time: the exponential distribution (together with its discrete counterpart, the geometric distribution) is the only one which enjoys being *memoryless*, that is, the probability of observing another n' errors after having seen n errors already is independent of n, or formally [33],

$$\Pr\left\{N > n + n' | N > n\right\} = \Pr\left\{N > n'\right\}.$$

Since the inter-arrival times between two errors are exponentially distributed with mean λ^{-1}, they do exhibit the memoryless property, which may appear unnatural in our setting when considering the physical sources of error. On the bright side, however, we explicitly know the expected number of errors within time span Δt, which is $\lambda \cdot (\Delta t)$. Assuming that the bits come in at frequency f, then if n denotes the number of arriving bits within Δt, we have $f \cdot \Delta t = n$ and the block-size in terms of bits is the solution of the equation $\lambda n / f = 1$ and thus comes to

$$\text{initial block-size } n \approx \frac{f}{\lambda}, \tag{4.4}$$

which is the block-size (in bits) that the Poisson process gives us. Observe that we have a constant block-size again, with the only difference to the previous Cascade variant being it originating from a probabilistic model rather than a time series smoothing approach, because the maximum likelihood estimator for the intensity parameter of a Poisson distribution is precisely the arithmetic mean. This is in accordance with our previous assumption on the availability of time series representing the average number of errors in several executions of the protocol. Hence, it can be seen as justification of the original proposal of the developers of Cascade [2], but offers no real advantage unless we can cleverly adapt λ.

4.5.3.1 Simulation of a Poisson Process

Knowing that the inter-arrival times between errors are exponentially distributed with parameter λ makes the simulation of error patterns particularly easy for constant λ, since we only have to draw from the exponential distribution. This can be done efficiently using the *inversion method* [36] as follows:

1. Draw U from a uniform distribution over the unit interval.
2. Set the random sample $X := -\frac{1}{\lambda}\ln(U)$, then X has an exponential distribution with parameter λ.

Figure 4.3 shows an example of errors simulated with this procedure. The impression of uniform distribution of n events occurring within the interval $(0, T)$ at times $t_1, \ldots, t_n$ is indeed correct, as it can be shown that given $N(T) = n$ errors have occurred up to time T at times $t_1, \ldots, t_n$ have the same distribution as the order statistics corresponding to n independent random variables uniformly distributed on $(0, T)$ [38].

Fig. 4.3 Example of a Poisson process error pattern with constant error rate. *Vertical lines* mark errors observed over the horizontal time axis

Coming back to the criticism, considering the physical sources of errors in photon transmission makes a constant error rate with independent errors across adjacent intervals appear unnatural. A more natural model would consider a non-constant error-rate parameter $\lambda(t)$, in which case, however, we lose the exponential distribution between two errors. A Poisson process with time-dependent (but deterministic) intensity parameter $\lambda(t)$ is called *inhomogeneous* and becomes a *Cox process* if the intensity parameter is replaced by a random variable. This type of model is discussed in the next section.

Let us consider the case of a deterministic intensity parameter $\lambda(t)$ first. We have prepared the basic for this in Sect. 4.5.1. Suppose that the error rate has been approximated by a model $\lambda \in \text{span}(g_1, \ldots, g_m)$ of integrable base functions g_i and that this model provides a satisfactory approximation of the Poisson rates over time. Following [38], by setting

$$\mu(t) := \int_0^t \lambda(s)ds,$$

one can show that

$$\Pr\{N(t+s) - N(t) = n\} = \exp(-[\mu(t+s) - \mu(t)])\frac{[\mu(t+s) - \mu(t)]^n}{n!},$$

which is again a Poisson distribution with rates $\mu(t+s)-\mu(t) = \int_t^{t+s} \lambda(\tau)d\tau$. Similarly as for standard Poisson processes, one can further deduce that the probability of one counted error within the interval $(t, t + \Delta t)$ is

$$\Pr\{\text{one counted error within}(t, t + \Delta t)\} = \lambda(t)\Delta t + o(\Delta t),$$

so as before, we may approximate the block-size in bits from the time span Δt according to the solution of $\lambda(t)\Delta t = 1$, which by calling f the bit frequency again (see above) gives

$$\text{block-size} \approx \frac{f}{\lambda(t)},$$

neglecting the terms $o(\Delta t)$.

4.5.3.2 Simulation of an Inhomogeneous Poisson Process

We follow in our presentation the lines of [22], remarking that the underlying method is better known as *rejection method* (cf. [36]). Let W be a bounded set over which we wish to simulate an inhomogeneous Poisson process. In our case, this will be the output bit string of the photon detection devices. In the first step of the simulation, a standard (homogeneous) Poisson process is simulated with constant intensity parameter $\lambda^* = \max_{t \in W} \lambda(t)$, where we assume λ^* finite. This is trivially satisfied in our setup if the base functions for $\lambda(t)$ are continuous and the length of the bit string is finite, so W is compact. The number of points in this pattern will be higher than the number in the final pattern, which arises by independent location-dependent thinning based on the thinning function $p(t) = \lambda(t)/\lambda^*$ [38]. The decision about whether or not a point from the first simulation is retained in the final output is made by taking $p(t)$ as its survival probability. In other words, a point x_n is retained with probability $p(x_n)$ independently of what happens to the other points. If the function $\lambda(t)$ is such that its maximum is difficult to determine, then any larger value may be used instead, resulting in a less efficient procedure, however, as more points will be rejected in that case. Figure 4.4 shows a simulated error pattern for an inhomogeneous point process with linear trend $\lambda(t) = at$ for some $a > 0$.

4.5.4 Using Bayesian Networks and Cox Processes

Cox processes may be considered as a generalization of inhomogeneous Poisson processes in the sense of having their intensity parameter stochastic rather than deterministic. For that reason, they are sometimes termed *doubly stochastic Poisson processes* [19, 22], and the first task for setting up a Cox process model will be finding a probabilistic model for the random intensity parameter Λ. Since errors in our model are due to random distortions, it appears natural to consider *temperature*,

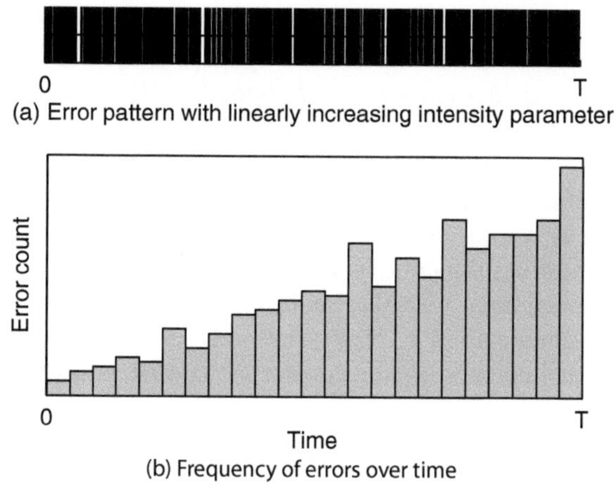

Fig. 4.4 Inhomogeneous Poisson process error pattern with linear trend. The *histogram* displays the number of errors per time-slot, showing a (linear) increase of errors over time

humidity, *vibration*, etc., as natural factors having influence on the error rate, as discussed in Chap. 7. For a concrete link, one may be able to quantify a number of additional aspects being relevant for the error rate, and an intuitive modeling technique for setting up joint probability distributions with known dependency relations among them is offered by the field of *probabilistic networks* (we refer the reader to [4, 10, 26, 32] and references therein for details). Call $Z_1, \ldots, Z_k$ a set of random variables representing environmental conditions on which the error rate λ depends on. A statistical analysis of the influence of temperature (Z_1), humidity (Z_2), as well as other environmental conditions, is provided in Chap. 7; see also [27]. Among these, dependency relations may exist which are either known or unknown. Even if they are unknown, a variety of learning strategies exist, taking existing data for inferring independency statements. Having set up a graph with arcs representing dependency between the nodes $Z_1, \ldots, Z_k$, one can sample from the so-modeled joint probability distributions using a variety of different techniques, as shown in the cited literature. The appeal of this modeling approach is its ability to capture the situation at hand almost directly, as we can translate several factors that have been identified to be of importance into a model using a widely automated procedure. It should be noticed, however, that many of the existing learning procedures for Bayesian networks (see [21] for instance), as well as inference procedures are computationally hard [9, 11, 12], even NP-complete [8]. Therefore, heuristic procedures for tackling this problem have been developed; see [25]. A restriction that applies to Bayesian network modeling is the acyclicity of graph models, meaning that no circular dependencies are allowed to exist like $Z_1 \to Z_2 \to Z_3 \to Z_1$ for instance. Relaxing the acyclicity assumption yields *dependency networks* [20]. If the directionality of the dependency is unknown, then an undirected link may be chosen, in which case we are dealing with *chain graphs* [7] if the network contains no partially directed cycles or with *Markov networks* if all links are undirected. Each of these comes with a comprehensive theory of learning its structure and doing inference.

As an example, consider temperature T, humidity H, and vibration V as influence factors, then a simple Bayesian network modeling the joint distribution of all four variables (i.e., including the error density f_Λ) is depicted in Fig. 4.5. Although the specification of the given Bayesian network requires knowledge of the distributions of T, H, V, as well as the distribution of Λ conditional on T, H, V, its real power lies in its learning capabilities. In other words, the error distribution conditional on the identified input variables can be updated during repeated protocol executions upon measurements of T, H, V. This adds further flexibility to our model, as repeated Bayesian updates of an estimator are consistent under some regularity conditions [1, 13].

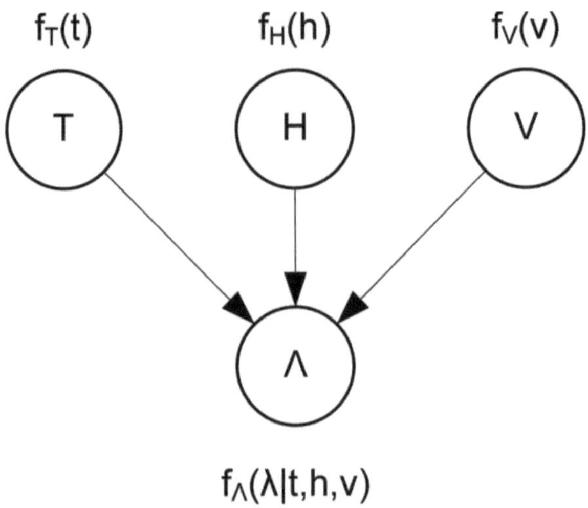

Fig. 4.5 Example Bayesian network model for random error rate Λ. Nodes are random variables with (conditional) densities assigned. Dependencies are modeled with directed arcs

The probabilistic network shown in Fig. 4.5 is a model for a simple random variable, whereas in our setting we consider the error rate to undergo seasonal variations (where the term *seasonal* may be understood as short- as well as long-duration changes). Consequently, we may set up a Poisson process with intensity parameter distributed according to a density induced by a Bayesian network.

A *Cox point process model* is set up in two steps [22]:

- Modeling of a stationary non-negative-valued stochastic process $\Lambda(t)$.
- Given a realization of the process $\Lambda(t)$ defining the local error rates over time, the values corresponding to this realization form an inhomogeneous Poisson process with intensity parameter $\Lambda(t)$.

4.5.4.1 Simulating Errors with Cox Processes

Simulation of a Cox process is straightforward by drawing a realization from $\Lambda(t)$ and simulating an inhomogeneous Poisson process with this time-dependent error rate.

4.6 Examples

Let us compare some of our proposals in terms of quality of the block-size selection strategy. Assume that the errors are scattered across a bit string, following an inhomogeneous Poisson distribution with an unknown intensity function roughly[1] having the shape depicted in Fig. 4.6. The horizontal axis covers a duration of $[0, T]$, one may think of the shape reflecting periods of high and low error rates, induced by vibrations due to traffic peak hours if T is set to 24 h. Extending the example further by including seasonal temperature variations or similar is left to the reader.

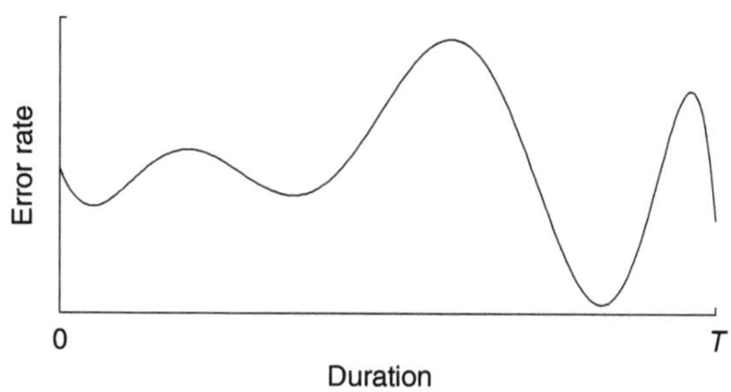

Fig. 4.6 Temporal variations of error rate (example)

Cascade with adaptive but a priori fixed initial block-sizes takes a constant block-size according to the median of the last w error rates, and take the block-size reciprocal to the (median-)estimated intensity parameter λ, according to expression (4.4), where the rate for one protocol execution is estimated in the maximum likelihood sense as

$$\hat{\lambda} = \frac{\text{number of errors}}{\text{total number of bits}}. \tag{4.5}$$

Clearly, taking constant block-sizes in the Cascade protocol implicitly assumes constant error rates over the whole duration of the protocol, so the Cascade variant with fixed initial block-size is best suited for slow variations of λ, such as being the

[1] None of the examples provided here is related to a real-life quantum cryptographic system, in order to adhere to intellectual property rights of the related data sets. Hence, all diagrams are given without explicit scales. All the data that have been used for the experiments described here come from simulations based on real-life system information.

case if we take $T = 24\,\text{h}$ in Fig. 4.6. We simulate 500 consecutive executions of the Cascade protocol over the duration $[0, T]$ with error rates varying as shown in Fig. 4.6 and a sliding median smoothing with window size $w = 11$.

In each execution, the final string is randomly changed (bit-flip to introduce an error) according to the (true) error pattern induced by the Poisson process, with intensity parameter varying according to the function in Fig. 4.6. Taking the median over the last 11 estimators given by (4.5), one can determine the (constant) block-size and count the number of errors falling into each block along the full length of the output string. Counting the number of blocks with $1, 2, 3, \ldots$ errors, we end up with a histogram, which for our example looks like shown in Fig. 4.7 (notice that the absolute numbers were normalized to represent empirical probabilities of encountering blocks with $1, 2, 3, \ldots$ errors).

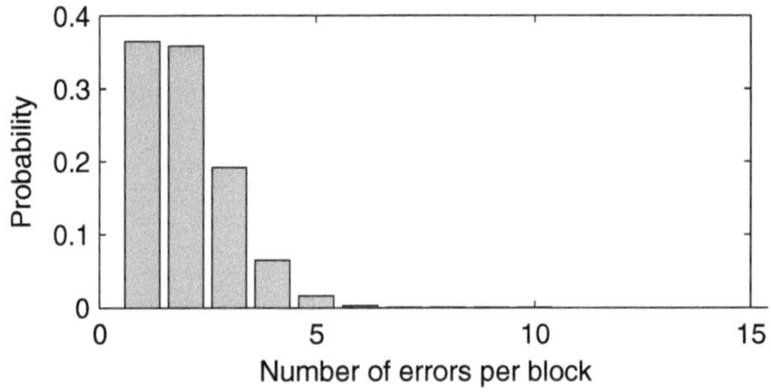

Fig. 4.7 Example probabilities for blocks with $1, 2, 3, \ldots$ errors for Cascade error correction with fixed initial block-size

For evaluating the dynamic variant of the protocol, let the function in Fig. 4.6 cover the duration of one single protocol execution within $[0, T]$. Simulation of an inhomogeneous Poisson process according to the rejection method sketched above and calculating the average number of errors in equidistantly distributed bins across the period $[0, T]$ give a set of points for which an approximating function $\Lambda(t)$ is sought. Figure 4.8 shows an example where this has been done and a polynomial model (solid line) has been fitted. Figure 4.9 displays the empirical probability of encountering blocks with $1, 2, 3, \ldots$ errors, showing that the variations in the local error rates can indeed be smoothed by our proposal. For comparison, Fig. 4.10 displays the distribution of the number of errors within blocks if a constant block-size is used. In the example, we took the maximum likelihood estimator from the simulated inhomogeneous Poisson process and derived the constant block-size using formula (4.4).

4.7 Summary

Our results show that adapting the block-size to variations of the local error rate is indeed worthwhile, since it significantly increases the efficiency of the error correction by reducing the number of bits that become revealed to the adversary

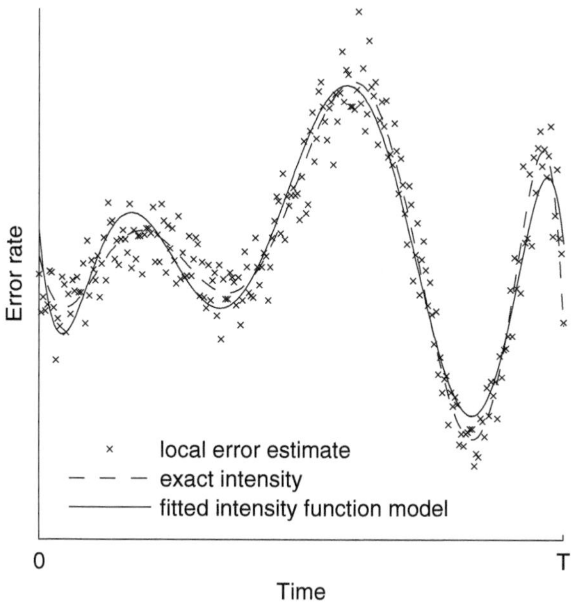

Fig. 4.8 Example of estimating local error rates and fitting a model to the empirical data

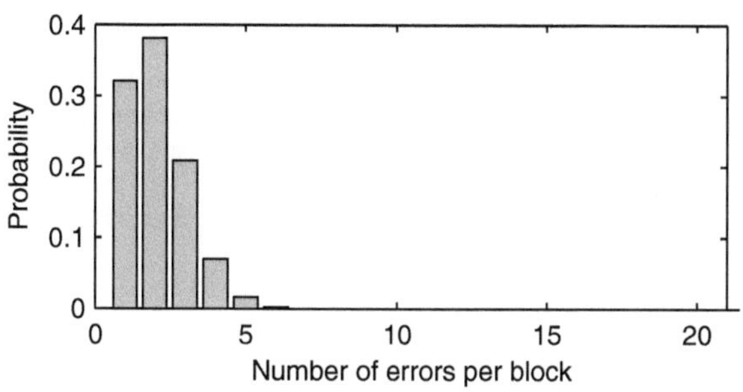

Fig. 4.9 Example empirical probabilities for blocks with $1, 2, 3, \ldots$ errors for Cascade with dynamic initial block-size

during the Cascade protocol. This leads to a considerable improvement of the QKD in terms of efficiency.

Combining the described models is possible by mixing a deterministic trend with a stochastic variational part to capture regular and predictable variations of the error rate which undergoes random perturbations through environmental influences. This can be seen as an extension to the least-squares model we proposed in Sect. 4.5.1, whereas it is known that the least-squares fit is a best linear unbiased estimator (BLUE; this is the Gauss–Markov theorem [23]) and becomes a best unbiased estimator (BUE) if the data enjoy a normal distribution. Hence, a combined approach is most suitable if the error is non-Gaussian, in which case a Bayesian model may be a promising alternative. A different view on the problem of partitioning a string into blocks such that the number of errors per block is minimized may also open the area

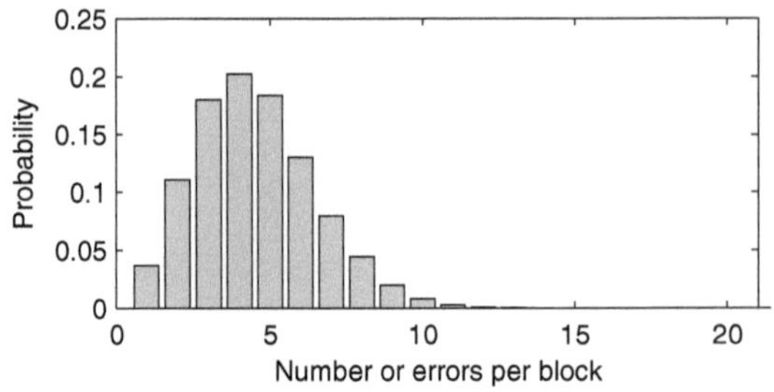

Fig. 4.10 Example empirical probabilities for blocks with $1, 2, 3, \ldots$ errors for Cascade with fixed initial block-size but strongly varying local error rates

for techniques known from *designs of experiments*. See [29, 35] for an introduction and overview.

References

1. Barron, A., Schervish, M.J., Wasserman, L.: The consistency of posterior distributions in nonparametric problems. Ann. Stat. **27**(2), 536–561 (1999)
2. Bennett, C.H., Bessette, F., Brassard, G., Salvail, L., Smolin, J.: Experimental quantum cryptography. J. Cryptol. **5**, 3–28 (1992)
3. Bollerslev, T., Engle, R.F., Nelson, D.B.: Handbook of Econometrics, Vol. IV, Chap. 49: ARCH Models, pp. 2959–3038. Elsevier Science B.V., Amsterdam (1994)
4. Borgelt, C., Kruse, R.: Graphical Models – Methods for Data Analysis and Mining. John Wiley & Sons, UK (2002)
5. Brassard, G., Salvail, L.: Secret-key reconciliation by public discussion. In: Heile-Beth, T (ed.) EUROCRYPT. Springer, New York, pp. 410–423 (1993)
6. Brockwell, P.J., Davis, R.A.: Introduction to Time Series and Forecasting. Springer, New York (1996)
7. Buntine, W.L.: Chain graphs for learning. In: Besnard, P. and Hanks, S. (eds.) Uncertainty in Artificial Intelligence, Morgan Kaufmann, San Francisco, CA., pp. 46–54 (1995)
8. Chickering, D.M.: Learning bayesian networks is NP-complete. In: D. Fisher, H.J. Lenz (eds.) Learning from Data: Artificial Intelligence and Statistics V, Chap. 12, pp. 121–130. Springer-Verlag New York (1996)
9. Cooper, G.F.: The computational complexity of probabilistic inference using bayesian belief networks (research note). Artif. Intell. **42**(2–3), 393–405 (1990)
10. Cowell, R.G., Dawid, A.P., Lauritzen, S.L., Spiegelhalter, D.J.: Probabilistic Networks and Expert Systems. Springer, New York (1999)
11. Dagum, P., Chavez, R.M.: Approximating probabilistic inference in bayesian belief networks. IEEE Trans. Pattern Anal. Mach. Intell. **15**(3), 246–255 (1993). DOI http://dx.doi.org/10.1109/34.204906
12. Dagum, P., Luby, M.: Approximating probabilistic inference in bayesian belief networks is NP-hard. Artif. Intell. **60**(1), 141–153 (1993)
13. Diaconis, P., Freedman, D.: On the consistency of bayes estimates. Ann. Stat. **14**(1), 1–26 (1986)
14. Dobson, A.J.: An introduction to generalized linear models, 2nd edn. Chapman & Hall, CRC (2002)
15. Doucet, A. (ed.): Sequential Monte Carlo Methods in Practice. Springer, New York (2001)

16. Faraway, J.J.: Extending the Linear Model with R. Chapman & Hall/CRC, Boca Ration (2006)
17. Gilbert, G., Hamrick, M.: Practical quantum cryptography: A comprehensive analysis (part one) (2000). URL http://www.citebase.org/abstract?id=oai:arXiv.org:quant-ph/00%09027
18. Gouriéroux, C.: ARCH Models and Financial Applications. Springer, New York (1997)
19. Grandell, J.: Doubly Stochastic Poisson Processes. Springer, New York (1976)
20. Heckerman, D., Chickering, D.M., Meek, C., Rounthwaite, R., Myers Kadie, C.: Dependency networks for inference, collaborative filtering, and data visualization. J. Mach. Learn. Res. **1**, 49–75 (2000)
21. Heckerman, D., Geiger, D., Chickering, D.M.: Learning bayesian networks: The combination of knowledge and statistical data. In: KDD Workshop, pp. 85–96 (1994)
22. Illian, J., Penttinen, A., Stoyan, H., Stoyan, D.: Statistical Analysis and Modeling of Spatial Point Patterns. Wiley, Chichestor (2008)
23. Kariya, T., Kurata, H.: Generalized Least Squares. Wiley, Chichestor (2004)
24. Kingman, J.: Poisson Processes. Oxford Science Publications, Oxford, UK (1993)
25. Larrañaga, P., Poza, M., Murga, R., Kuijpers, C.: Structure learning of bayesian networks by genetic algorithms: A performance analysis of control parameters. IEEE J. Pattern An. Mach. Intell. **18**(9), 912–926 (1996)
26. Lauritzen, S.L.: Graphical Models. Oxford Statistical Science Series 17. Oxford Science Publications, New York (1996)
27. Lessiak, K., Kollmitzer, C., Schauer, S., Pilz, J., Rass, S.: Statistical analysis of QKD networks in real-life environments. In: Proceedings of the Third International Conference on Quantum, Nano and Micro Technologies. IEEE Computer Society (2009) 109–114
28. Lindsey, J.K.: Applying Generalized Linear Models. Springer, New York (1997)
29. Mason, R.L.: Statistical Design and Analysis of Experiments with Applications to Engineering and Science. Series in Probability and Statistics. Wiley, New York (2003)
30. McCullagh, P., Nelder, J.: Generalized Linear Models, 2nd edn. Monographs on Statistics and Applied Probability 37. Chapman & Hall, London (1989)
31. Pankratz, A.: Forecasting with Univariate Box-Jenkins Models. Wiley, New York (1983)
32. Pearl, J.: Probabilistic Reasoning in Intelligent Systems: Networks of Plausible Inference. Morgan Kaufmann Publishers Inc., San Francisco, CA (1988)
33. Pitman, J.: Probability. Springer, New York (2000)
34. Press, W.H., Teukolsky, S.A., Vetterling, W.T., Flannery, B.P.: Numerical Recipes in C, 2nd edn. Cambridge University Press, New York (1992)
35. Rasch, D., Herrendörfer, G., Bock, J., Busch, K.: Verfahrensbibliothek, Versuchsplanung und -auswertung. Deutscher Landwirtschaftsverlag, Berlin (1978)
36. Ripley, B.D.: Stochastic Simulation. Wiley, New York (1987)
37. Robert, C.P.: The Bayesian Choice. Springer-Verlag, New York (2001)
38. Ross, S.M.: Stochastic Processes. Series in Probability and Mathematical Statistics. Wiley, New York (1983)
39. Sherman, J., Morrison, W.J.: Adjustment of an inverse matrix corresponding to a change in one element of a given matrix. Ann. Math. Statistics **21**, 124–127 (1950)
40. Štulajter, F.: Predictions in Time Series Using Regression Models. Springer, New York (2002)

Chapter 5
Attack Strategies on QKD Protocols

S. Schauer

In the following we will describe a number of different attack strategies on Quantum Key Distribution protocols. First, we will discuss QKD protocols based on ideal sources, mainly using single photons as in [2, 1, 11] but also entanglement-based protocols as, for example, in [15]. Additionally, we will deal with the security of QKD protocols in realistic environments as they are described in Chap. 6. Regarding realistic sources, due to the physical limitations of the apparatus some loopholes were identified, which could lead to a successful attack on such protocols. As we will see, special countermeasures are implemented in the protocols to close those loopholes and guarantee an unconditional secure communication.

The security discussions are limited to *individual attacks*, since they are the simplest and most studied class of attack strategies on QKD protocols. The key point of an individual attack is that the eavesdropper, Eve, interacts with each signal coming from Alice separately. Some strategies rely on the fact that Eve is able to postpone her interaction with the signal (i.e., the measurement) until after the sifting and error correction – or as long as she wants – to obtain the maximum amount of information from the public discussion. Alternatively, Eve measures the state instantly and uses the information from sifting and error correction later on.

The individual attacks stand in opposition to the *collective attacks*, where Eve prepares an ancilla state for each signal coming from Alice and lets it interact with the signal. Eve passes the signal on to Bob and keeps the ancilla at her side. When all signals are transferred Eve waits to obtain as much information as possible from the public communication to perform the best measurement on all of her ancillae. It has been proven that the same security bound defined for QKD protocols also holds if collective attacks are applied [8, 7].

The most general version of collective attacks are *coherent attacks*. In this kind of attack Eve is allowed to perform any quantum operation on the signal in transit and use any possible ancilla state. Due to this vast amount of possibilities for Eve the attack cannot be parametrized and brute force optimization is not possible. Anyhow,

---

S. Schauer (✉)
Safety & Security Department, Quantum Technologies, AIT Austrian Institute of Technology GmbH, Lakeside B01A, 9020 Klagenfurt, Austria, `stefan.schauer@ait.ac.at`; `http://www.ait.ac.at`

Schauer, S.: *Attack Strategies on QKD Protocols*. Lect. Notes Phys. **797**, 71–95 (2010)
DOI 10.1007/978-3-642-04831-9_5 © Springer-Verlag Berlin Heidelberg 2010

bounds for information and theoretical security have been found which are equal to the bounds for collective attacks [34].

5.1 Introduction

In the discussions about the security of QKD protocols the amount of information Eve has about the secret key shared between Alice and Bob is of great interest. Every attack on a QKD protocol is characterized by this amount of information and Alice's and Bob's goal is to minimize it. Usually, Eve's information about the secret key consists of two parts: the information she obtains from her measurement on the signal in transit and the information about Alice's and Bob's choice of bases. The second part tells Eve whether she made a correct measurement for the respective qubit or not. The best way to express this fact is to use the conditional probability $p(s|m)$. Here, s is the value of the bit Alice sent and m is Eve's measurement outcome. The value $p(s|m)$ can easily be computed using the probabilities $p(m|s)$, i.e., the probability that Eve obtains the result m if Alice originally sent the bit s. To achieve that the following formula is used:

$$p(s|m) = \frac{p(m|s)}{\sum_{s'} p(m|s')}. \tag{5.1}$$

The next interesting value is the probability that Eve obtains the same result as Alice, which is the *collision probability*

$$P_c(s|m) = \sum_s p(s|m)^2. \tag{5.2}$$

Further, the expected collision probability over all possible measurement outcomes m of Eve is described as

$$\langle P_c \rangle = \sum_m p(m) P_c(s|m). \tag{5.3}$$

To quantify Eve's amount of information on Alice's bit two entropies are used: the *Shannon entropy* H and the *Renyi entropy* R (cf. also "Sect. 2.3"). Since the information depends on Eve's measurement result, the conditional versions of these two entropies are used, i.e., $H(S|M)$ and $R(S|M)$, respectively. The Renyi entropy can be computed directly from the collision probability $P_c(s|m)$

$$R(S|M = m) = -\log P_c(s|m) = -\log \sum_s p(s|m)^2 \tag{5.4}$$

and is averaged over the respective probabilities of Eve's measurement results, i.e.,

$$R(S|M) = \sum_m p(m) R(S|M = m) \tag{5.5}$$

Similarly, the Shannon entropy is defined as

$$H(S|M=m) = -\sum_{s} p(s|m)\log p(s|m) \tag{5.6}$$

and also averaged over the probabilities of Eve's results

$$H(S|M) = \sum_{m} p(m)H(S|M=m). \tag{5.7}$$

The Shannon entropy H is an estimator of the uncertainty of a probability distribution and thus the variation of the Shannon entropy can be interpreted as the information gain. For the a priori probability distribution X and the a posteriori distribution Y the information gain is $I = H(X) - H(Y)$. This can be used to describe the amount of information Eve obtains by applying a specific attack. Here Eve has no a priori information about the secret key and thus $H = 1$, since Alice chooses her bit string at random. Therefore, the amount of information gained by Eve is $I = 1 - H(S|M)$, which will be used constantly in the following sections.

Another important question is how much key material has to be discarded to minimize Eve's knowledge about the key. This amount is called the *discarded fraction* τ and is computed using the expected collision probability

$$\tau = 1 + \log\langle P_c\rangle^{\frac{1}{n}}. \tag{5.8}$$

Following from this equation a bit string of length n has to be reduced by $n\tau$ bits during privacy amplification to leave Eve with at most 1 Shannon bit of information on the whole secret key, no matter its length [24].

5.2 Attack Strategies in an Ideal Environment

5.2.1 Intercept and Resend

5.2.1.1 Naive Intercept and Resend

The most intuitive kind of an individual attack is the *intercept and resend* (*I&R*) attack. The main intention for Eve is to get hold of each photon coming from Alice and measuring it in some predefined basis. According to her result Eve prepares a new photon and forwards it to the legitimate receiver, Bob. In detail, Alice's qubit will either be in the horizontal/vertical or the diagonal basis, explicitly one of the four states $|H\rangle$, $|V\rangle$, $|+\rangle$, or $|-\rangle$, where

$$|+\rangle = \frac{1}{\sqrt{2}}\big(|H\rangle + |V\rangle\big) \qquad |-\rangle = \frac{1}{\sqrt{2}}\big(|H\rangle - |V\rangle\big). \tag{5.9}$$

If Alice sends a 0, she will either encode it into $|H\rangle$ or $|+\rangle$ with equal probability. Eve, unaware of Alice's encoding, will choose randomly between the H/V and the $+/-$ basis. Thus, she will obtain a correct result in case Alice sent $|H\rangle$ and Eve measured in the H/V basis or Alice sent $|+\rangle$ and Eve measured in the $+/-$ basis, respectively. Any other combination will result in a completely random measurement outcome. This leads to the scheme in Fig. 5.1.

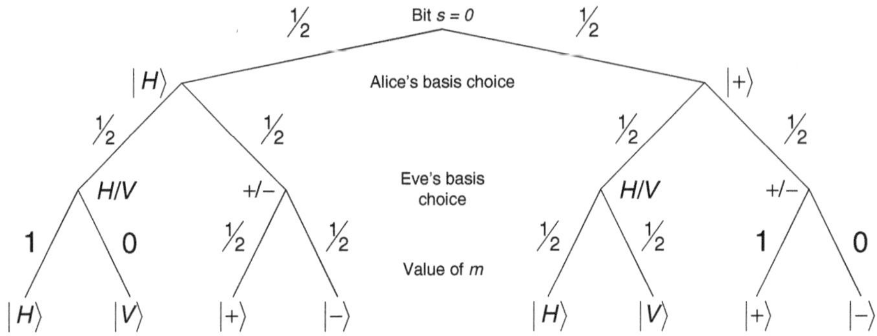

Fig. 5.1 Decision tree for the naive I&R attack strategy

For now, we will also assume that Eve does not listen to any public communication between Alice and Bob. Therefore, she will not know in which case her measurement was wrong. The conditional probabilities $p(m|s)$ for the four possible results are then

$$\begin{aligned} p\big(m = |H\rangle|s = 0\big) = p\big(m = |+\rangle|s = 0\big) &= \left(\frac{1}{2}\right)^3 + \left(\frac{1}{2}\right)^2 \cdot 1 = \frac{3}{8} \\ p\big(m = |V\rangle|s = 0\big) = p\big(m = |-\rangle|s = 0\big) &= \left(\frac{1}{2}\right)^3 + \left(\frac{1}{2}\right)^2 \cdot 0 = \frac{1}{8} \end{aligned} \tag{5.10}$$

and equally for $p(m|s = 1)$. For the conditional probabilities $p(s|m)$ the sum $\sum_s p(m|s) = \frac{1}{2}$ and thus we get $p(s|m) = 2p(m|s)$. Further, the collision probability in the naive version of the I&R attack is

$$P_c\big(s|m = |H\rangle\big) = \left(\frac{3}{4}\right)^2 + \left(\frac{1}{4}\right)^2 = \frac{5}{8} \tag{5.11}$$

and similarly for $m = |V\rangle$, $m = |+\rangle$, and $m = |-\rangle$ which results in the average collision probability

$$\langle P_c\rangle = \sum_m \frac{1}{4} P_c(s|m) = 4\left(\frac{1}{4}\right)\left[\left(\frac{1}{4}\right)^2 + \left(\frac{3}{4}\right)^2\right] = \frac{5}{8}. \tag{5.12}$$

From the collision probability the discarded fraction can be computed as $1 - \log \langle P_c \rangle$ which is $\tau \simeq 0.322$. Thus, only one-third of the key has to be discarded to guarantee that Eve has less than one bit of information on the whole key.

Looking at the Renyi entropy for $m = |H\rangle$, we get

$$R\big(S|M = |H\rangle\big) = -\log P_c\big(s|m = |H\rangle\big) = -\log \frac{5}{8} = 3 - \log 5. \tag{5.13}$$

Since the Renyi entropy for $m = |V\rangle$, $m = |+\rangle$, and $m = |-\rangle$ is the same and all four results are equally probable, the average Renyi entropy is

$$R\big(S|M\big) = \sum_m \frac{1}{4} R\big(S|M = m\big) = 4\left(\frac{1}{4}\right)\big(3 - \log 5\big) = 3 - \log 5. \tag{5.14}$$

The second interesting entropy is the conditional Shannon entropy $H(S|M = |H\rangle)$

$$H\big(S|M = |H\rangle\big) = -\frac{3}{4}\log\frac{3}{4} - \frac{1}{4}\log\frac{1}{4} = 0.811, \tag{5.15}$$

which is equal to the other entropies $H(S|M = |V\rangle)$, $H(S|M = |+\rangle)$, $H(S|M = |-\rangle)$, such that

$$H\big(S|M\big) = \sum_m \left(\frac{1}{4}\right) H\big(S|M = m\big) = 4\left(\frac{1}{4}\right)\left(-\frac{3}{4}\log\frac{3}{4} - \frac{1}{4}\log\frac{1}{4}\right) = 0.811. \tag{5.16}$$

The total information Eve will have in the end about each bit is $1 - H\big(S|M\big) \simeq 0.2$, which is rather poor for Eve. Therefore, Eve will use another strategy, which gives her more information. One possibility is to use another measurement basis, e.g., the *Breidbart basis.*

5.2.1.2 Intercept and Resend in the Breidbart Basis

As pointed out above, we did not take into account that Eve is able to listen to the public discussion of Alice and Bob during the error correcting and sifting stage in the naive I&R attack. If Eve listens to the public discussion she can obtain more information since she knows in which cases she did a correct measurement. Furthermore, for this attack strategy Eve restricts herself to only one measurement basis and gains more information compared to the naive I&R attack. The measurement basis in this attack is the $\varphi_\pm$ basis, which is

$$|\varphi_+\rangle = a|H\rangle + b|V\rangle \qquad |\varphi_-\rangle = b|H\rangle - a|V\rangle. \tag{5.17}$$

Here, $a = \sin\alpha$ and $b = \cos\alpha$ depend on some angle α. Using this basis for Eve's measurement we get a slightly different decision tree (cf. Fig. 5.2).

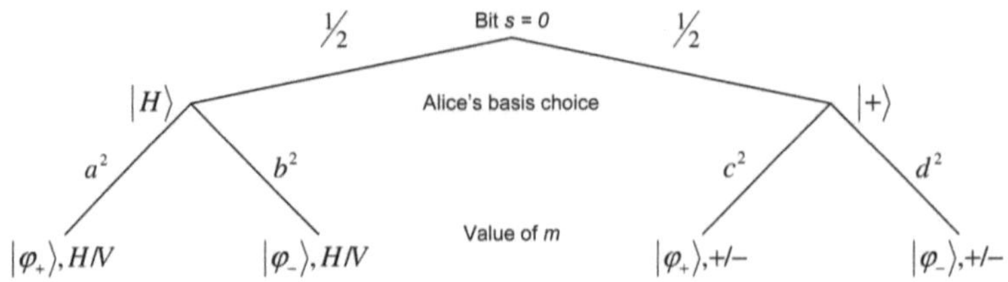

Fig. 5.2 Decision tree for the I&R attack strategy in the Breidbart basis

If Alice sends a 0 encoded as $|H\rangle$, Eve will obtain $|\varphi_+\rangle$ with probability a^2 and $|\varphi_-\rangle$ with probability b^2. If Alice encodes 0 as $|+\rangle$, Eve will obtain $|\varphi_+\rangle$ with probability $c^2 = \frac{1}{2}(a+b)^2$ and $|\varphi_-\rangle$ with probability $d^2 = \frac{1}{2}(a-b)^2$, which is a direct result from the measurement of $|+\rangle$ in the $|\varphi_\pm\rangle$-basis. This gives the conditional probabilities

$$\begin{aligned} p\Big(m = (|\varphi_+\rangle, H/V)|s=0\Big) &= \frac{1}{2}a^2 = p\Big(m = (|\varphi_-\rangle, H/V)|s=1\Big) \\ p\Big(m = (|\varphi_-\rangle, H/V)|s=0\Big) &= \frac{1}{2}b^2 = p\Big(m = (|\varphi_+\rangle, H/V)|s=1\Big) \\ p\Big(m = (|\varphi_+\rangle, +/-)|s=0\Big) &= \frac{1}{2}c^2 = p\Big(m = (|\varphi_-\rangle, -/+)|s=1\Big) \\ p\Big(m = (|\varphi_-\rangle, +/-)|s=0\Big) &= \frac{1}{2}d^2 = p\Big(m = (|\varphi_+\rangle, -/+)|s=1\Big) \end{aligned} . \qquad (5.18)$$

Since $a^2 + b^2 = c^2 + d^2 = 1$, the sum $\sum_s p(m|s) = \frac{1}{2}$ and thus the conditional probabilities are $p(s|m) = 2p(m|s)$. This gives a collision probability of

$$P_c\Big(s|m = (|\varphi_+\rangle, H/V)\Big) = \left(a^2\right)^2 + \left(b^2\right)^2 = a^4 + b^4, \qquad (5.19)$$

which leads to the average collision probability (using some simple algebra)

$$\langle P_c \rangle = \frac{1}{4}\left(2a^4 + 2b^4\right) + \frac{1}{4}\left(2c^4 + 2d^4\right) = \frac{1}{2}\left(a^4 + b^4 + c^4 + d^4\right) = \frac{3}{4}. \qquad (5.20)$$

Based on this result, the discarded fraction τ can be computed as described in Eq. 5.8 which leads to $\tau \simeq 0.585$. Thus, more than half of the bits of the secret key have to be discarded to minimize Eve's information to at most one bit.

It has to be stressed that the average collision probability is independent of the angle α chosen by Eve. Similarly, also the Renyi entropy

$$R\Big(S|M = (|\varphi_+\rangle, H/V)\Big) = -\log(a^4 + b^4) \qquad (5.21)$$

and consequently the average Renyi entropy

$$\begin{aligned} R(S|M) &= \frac{1}{2}(a^2+b^2) - \log(a^4+b^4) + \frac{1}{2}(c^2+d^2) - \log(c^4+d^4) \\ &= 2 - \log 3 \end{aligned} \tag{5.22}$$

is independent of the angle α. This is not the case for the Shannon entropy, since

$$H\Big(S|M = \big(|\varphi_+\rangle, H/V\big)\Big) = -a^2 \log a^2 - b^2 \log b^2 \tag{5.23}$$

and further

$$\begin{aligned} H(S|M) &= \frac{1}{2}(a^2+b^2)(-a^2 \log a^2 - b^2 \log b^2) \\ &\quad + \frac{1}{2}(c^2+d^2)(-c^2 \log c^2 - d^2 \log d^2) \\ &= -a^2 \log a^2 - b^2 \log b^2. \end{aligned} \tag{5.24}$$

To make Eve's errors in the H/V and $+/-$ basis equally probable she uses $\alpha = \frac{3\pi}{8}$. In this case the basis $\varphi_\pm$ is called the *Breidbart basis*, as introduced in [3]. As pointed out above, the average collision probability and the Renyi entropy are independent of the angle α and it only affects the Shannon entropy, which is in that case

$$H(S|M) = -\sin^2\frac{3\pi}{8}\log_2\sin^2\frac{3\pi}{8} - \cos^2\frac{3\pi}{8}\log_2\cos^2\frac{3\pi}{8} = 0,61. \tag{5.25}$$

Thus, Eve learns $1 - H(S|M) \simeq 0.4$ bits from every bit sent by Alice. Compared to the intuitive and rather simple approach described above Eve gains much more information with a measurement in the Breidbart basis (cf. Eq. 5.16). Nevertheless, the amount of information is not yet optimal. As we will see in the next section, Eve is able to increase her information about Alice's bits by using the H/V and $+/-$ bases again and listening to the communication between Alice and Bob.

5.2.1.3 Full Intercept and Resend

The most successful I&R attack is a mixture of the two aforementioned strategies. In this attack Eve randomly chooses between the H/V and the $+/-$ basis to measure the signals coming from Alice. She forwards the results she obtains to Bob and listens to the public communication between Alice and Bob during the sifting phase. If Alice sends a 0 encoded as $|H\rangle$, Eve will either measure it in the H/V or $+/-$ basis. As we have already seen, if Eve uses the H/V basis she will obtain $|H\rangle$ with certainty and introduce no error. Otherwise, she will obtain $|+\rangle$ or $|-\rangle$ with equal probability (cf. Fig. 5.3).

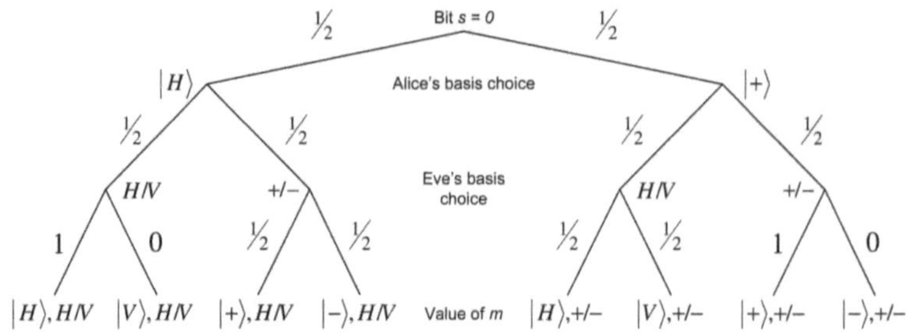

Fig. 5.3 Decision tree for the full I&R attack strategy

Comparing the decision tree with the one from the naive I&R attack, it is easy to see that Eve can eliminate two events for $s = 0$, i.e., if she measured $|V\rangle$ and Alice used the H/V basis and $|-\rangle$ and Alice used the $+/-$ basis. These two events occur with probability $p = 0$ which increases Eve's information. In detail, the probabilities $p(m|s)$ are

$$\begin{aligned}
p\Big(m = (|H\rangle, H/V)|s = 0\Big) &= \left(\frac{1}{2}\right)^2 \cdot 1 = \frac{1}{4} = p\Big(m = (|+\rangle, +/-)|s = 0\Big)\\
p\Big(m = (|V\rangle, H/V)|s = 0\Big) &= \left(\frac{1}{2}\right)^2 \cdot 0 = 0 = p\Big(m = (|-\rangle, +/-)|s = 0\Big)\\
p\Big(m = (|+\rangle, H/V)|s = 0\Big) &= \left(\frac{1}{2}\right)^3 = \frac{1}{8} = p\Big(m = (|H\rangle, +/-)|s = 0\Big)\\
p\Big(m = (|-\rangle, H/V)|s = 0\Big) &= \left(\frac{1}{2}\right)^3 = \frac{1}{8} = p\Big(m = (|V\rangle, +/-)|s = 0\Big)
\end{aligned}, \tag{5.26}$$

and we get similar values for $s = 1$. For the sum $\sum_s p(m|s)$ we obtain $\frac{1}{4}$, such that $p(s|m) = 4p(m|s)$. This results in the collision probabilities 1 if Eve chooses the correct basis for her measurement and $\frac{1}{2}$ if she chooses a basis different from Alice's preparation. Thus, the average collision probability is

$$\langle P_c \rangle = \frac{1}{4} + 4\frac{1}{16} + \frac{1}{4} = \frac{3}{4}. \tag{5.27}$$

Calculating the discarded fraction τ, we get $\tau \simeq 0.585$ as for the attack in the Breidbart basis (cf. Sect. 5.2.1.2).

For the Renyi entropy we obtain $R(S|M = m) = -\log 1 = 0$ whenever Eve's choice of the basis is correct and $R(S|M = m) = -\log 0.5 = 1$ otherwise. The average Renyi entropy is then

$$R(S|M) = \frac{1}{2}(0 + 1) = \frac{1}{2}. \tag{5.28}$$

For the Shannon entropy we also get either 0, if Eve guessed the same basis as Alice and $\frac{1}{2}$ otherwise. This results in a Shannon entropy of

$$H\big(S|M\big) = 4\frac{1}{8} = \frac{1}{2}. \tag{5.29}$$

This strategy gives more information to Eve than the naive approach (cf. Eq. 5.16) or the I&R attack in the Breidbart basis (cf. Eq. 5.25).

5.2.2 *Entanglement-Based Attacks*

5.2.2.1 BB84-Like Protocols

Another strategy for Eve is to use entanglement to interact with the signal sent by Alice. In this case Eve prepares a probe for each signal, entangles the probe with it and then passes only the signal on to Bob. Later on, Eve is able to perform a measurement or any other quantum operation on the probe in her possession to gain information about the original signal. As pointed out above, the analysis in this chapter is restricted to individual attacks only. Thus, we will just look at scenarios where Eve performs her operation on one single probe. An operation on a subset or even all of the probes is a coherent measurement, as described above, which will not be discussed here.

Taking the BB84 protocol [2] which we also referred to in Sect. 5.2.1, a rather simple strategy for Eve is to use an entangled pair, i.e., one of the Bell states

$$|\Phi^p m\rangle = \frac{1}{\sqrt{2}}\Big(|00\rangle \pm |11\rangle\Big) \qquad |\Psi^p m\rangle = \frac{1}{\sqrt{2}}\Big(|01\rangle \pm |10\rangle\Big) \tag{5.30}$$

and to perform a measurement in the Bell basis on the photon coming from Alice together with one of the entangled photons. This is equal to a quantum teleportation scheme [4, 5, 20] where the unknown signal state is teleported onto Eve's probe

$$\begin{aligned}\big(\alpha|H\rangle + \beta|V\rangle\big)\frac{1}{\sqrt{2}}\Big(|HH\rangle + |VV\rangle\Big) = \\ \frac{1}{2}\Big(|\Phi^+\rangle\Big(\alpha|H\rangle + \beta|V\rangle\Big) + |\Phi^-\rangle\Big(\alpha|H\rangle - \beta|V\rangle\Big) \\ + |\Psi^+\rangle\Big(\alpha|V\rangle + \beta|H\rangle\Big) + |\Psi^-\rangle\Big(\alpha|V\rangle - \beta|H\rangle\Big)\Big).\end{aligned} \tag{5.31}$$

Eve is able to keep her probe until Alice reveals her basis choice and measure it in the correct basis to obtain full information. If we look at the average collision probability and Eve's Shannon information about Alice's bit we see that

$$\big\langle P_c\big\rangle = 1 \qquad \text{and} \qquad 1 - H\big(S|M\big) = 1. \tag{5.32}$$

Eve has full information about the bit Alice sent. Nevertheless, the signal, which Eve has forwarded to Bob is now in a Bell state, i.e., it has lost every information about Alice's basis choice and is in a completely mixed state. Bob will obtain a random result upon a measurement in the H/V as well as in the $+/-$ basis which is easy to see from Bob's average collision probability $\langle P_c \rangle = \frac{1}{2}$. Thus, Alice and Bob will detect too many errors during their sifting phase (around 50%) and therefore will abort the protocol.

As we see, regarding the BB84 protocol, Eve gains full information about Alice's bit using this attack strategy but the average collision probability is the same compared to the full I&R strategy (cf. Eq. 5.27). Thus, she can gain no additional information from this strategy.

But what if Alice and Bob use entangled states for communication as in the Ekert protocol [15]? A strategy for Eve in this case is to prepare her probes in the state $|H\rangle$ and perform a controlled NOT (CNOT) operation (cf. Eq. 5.33) on the signal and her probe

$$\text{CNOT}_{12} = |H\rangle\langle H| \otimes \mathbb{I} + |V\rangle\langle V| \otimes \Big(|0\rangle\langle 1| + |1\rangle\langle 0|\Big). \tag{5.33}$$

The CNOT operation is a quantum operator acting on two qubits, a source and a target. If the source is in the state $|V\rangle$, a NOT or flip operation is performed on the target qubit. The subscripts in Eq. 5.33 denote which qubit is the source (first index) and which is the target (second index). This is convenient if the CNOT operation is applied on a state consisting of more than two qubits to avoid confusions. The CNOT operation applied on the signal coming from Alice and Eve's probe will alter the state into

$$\text{CNOT}_{23}|\Phi^+\rangle \otimes |H\rangle = \frac{1}{\sqrt{2}}\Big(|HHH\rangle + |VVV\rangle\Big). \tag{5.34}$$

The resulting state is a GHZ state [17] which has the special property that if one of the photons is measured the other two photons immediately collapse into a certain state depending upon the measurement result. In case of Eq. 5.34 if Alice measures in the H/V basis, Bob and Eve will obtain the same result as Alice if they also perform their measurement in the H/V basis. In case Alice uses the $+/-$ basis, Bob's measurement result in the same basis will not correlate with Alice's result 50% of the times. For the collision probability and Shannon information this means

$$\big\langle P_c \big\rangle = 1 \qquad \text{and} \qquad 1 - H\big(S|M\big) = 1 \tag{5.35}$$

if Alice and Bob measure in the H/V basis. For a measurement in the $+/-$ basis, Bob obtains the same result as Alice with probability $\langle P_c \rangle = \frac{1}{2}$. Therefore, the overall information Eve obtains on each secret bit is $1 - H(S|M) = 0.75$, which is significantly more compared to the I&R strategies discussed in Sect. 5.2.1. Nevertheless, an error is detected with probability 0.5 every time Alice and Bob use the $+/-$ basis. This unbalanced occurrence of errors makes it easy for Alice and Bob to identify the presence of Eve.

5.2.2.2 ZLG Attack

Besides the Ekert protocol, there are other entanglement-based protocols which make use of a phenomenon of quantum mechanics called entanglement swapping [40, 9]. Entanglement swapping is a special case of quantum teleportation where a completely mixed state is teleported and thus the entanglement between certain photons is changed. An attack strategy based on entanglement swapping is a rather theoretical approach because the realization is very complex due to limitations of the physical apparatus. Nevertheless, attack strategies based on entanglement swapping have to be considered because Eve could have the physical means to perform such attacks and some protocols have already been shown to be insecure against them.

An example for a protocol open to an attack based on entanglement swapping is a protocol presented by Adan Cabello [13]. In this protocol Alice has two entangled pairs in the state $|\Psi^-\rangle_{12}$ and $|\Psi^+\rangle_{35}$, whereas Bob has one pair in the state $|\Psi^+\rangle_{46}$ (cf. (1) in Fig. 5.4). Alice sends qubit 2 to Bob and performs a Bell state measurement on qubits 1 and 3 in her possession which entangles qubits 2 and 5 due to entanglement swapping:

$$\begin{aligned}|\Psi^-\rangle_{12}|\Psi^+\rangle_{35} = \frac{1}{2}\Big(&|\Phi^-\rangle_{13}|\Phi^+\rangle_{25} - |\Phi^+\rangle_{13}|\Phi^-\rangle_{25} \\ &+ |\Psi^-\rangle_{13}|\Psi^+\rangle_{25} - |\Psi^+\rangle_{13}|\Psi^-\rangle_{25}\Big).\end{aligned} \tag{5.36}$$

In detail, if qubits 1 and 3 are in the state $|\Psi^-\rangle_{13}$ after the Bell state measurement, Alice knows that qubits 2 and 5 are in the state $|\Psi^+\rangle_{25}$ (cf. (2) and (3) in Fig. 5.4). After receiving qubit 2 from Alice, Bob also performs a Bell state measurement on qubits 2 and 4 and obtains, for example, $|\Phi^+\rangle_{24}$. Now, qubits 5 and 6 are in the entangled state $|\Phi^+\rangle_{56}$ (cf. Eq. 5.37 and (4) in Fig. 5.4).

$$\begin{aligned}|\Psi^+\rangle_{25}|\Psi^+\rangle_{46} = \frac{1}{2}\Big(&|\Phi^+\rangle_{24}|\Phi^+\rangle_{56} - |\Phi^-\rangle_{24}|\Phi^-\rangle_{56} \\ &+ |\Psi^+\rangle_{24}|\Psi^+\rangle_{56} - |\Psi^-\rangle_{24}|\Psi^-\rangle_{56}\Big).\end{aligned} \tag{5.37}$$

Bob sends qubit 6 to Alice, who is able to determine the state of qubits 5 and 6 by measuring them in the Bell basis. She publicly announces her result and both parties are able to calculate the state of qubits 1, 3 and 2, 4, respectively. Alice and Bob use these two states to agree upon a shared secret key.

In a comment [38] on the Cabello protocol Zhang, Li, and Guo presented an attack strategy which gives an adversary full information about the key shared between Alice and Bob (we will call this the *ZLG attack* further on). The idea is that Eve prepares an entangled pair $|\Psi^+\rangle_{78}$ and uses qubit 7 to replace qubit 2 flying to Bob (cf. (2) in Fig. 5.5). Since Bob's measurement on qubits 4 and 7 entangles qubits 6 and 8, Eve intercepts qubit 6 coming from Bob and performs a Bell state measurement on it together with qubit 8. As we have seen above, qubits 2 and 5 are in the state $|\Psi^+\rangle_{25}$ according to Alice's measurement. Based on the outcome

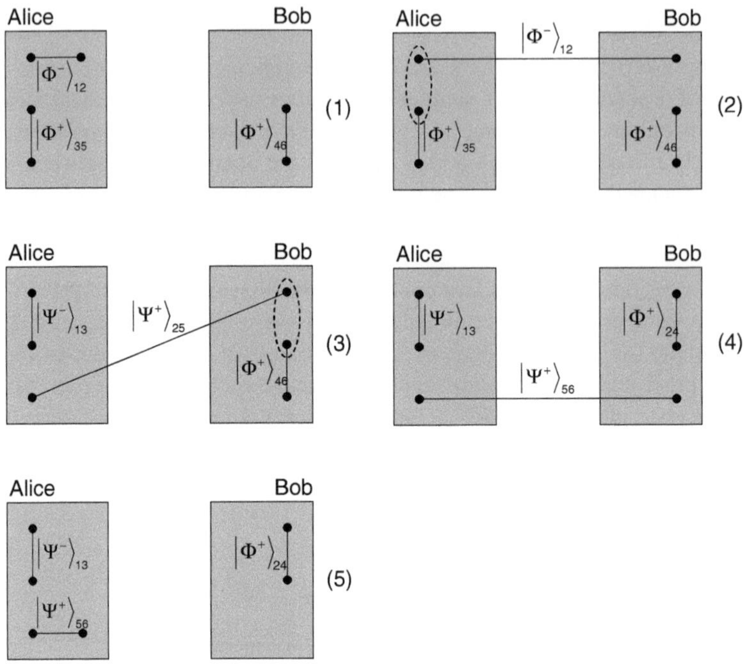

Fig. 5.4 Illustration of the protocol in [13]

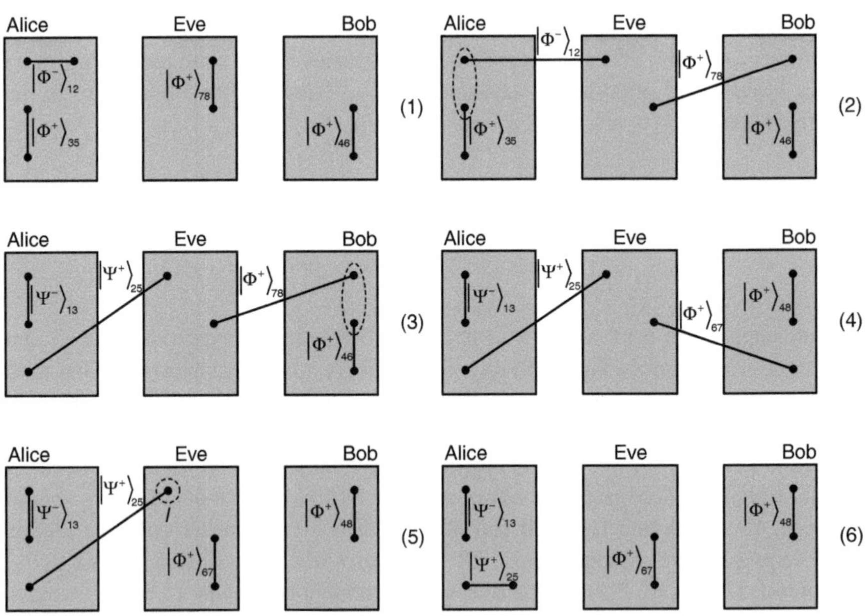

Fig. 5.5 Illustration of the ZLG attack scenario [38]

of her measurement, Eve knows the exact result of Bob's measurement (cf. (4) in Fig. 5.5). Moreover, she also knows how to change the state of qubits 2 and 5 such that the state of qubits 5 and 6 will correspond to Alice's and Bob's result. Therefore, Eve uses one of the Pauli operators $\mathbb{I}$, σ_x, σ_y, and σ_z onto qubit 2 to alter the state $|\Psi^+\rangle_{25}$ (cf. (5) in Fig. 5.5). When Eve returns qubit 2 to Alice, Alice performs her measurement and will obtain a result correlated to Bob's measurement outcome, as it would be expected (compare (5) in Fig. 5.4 and (6) in Fig. 5.5). Since Eve's qubits 6 and 8 are in the same state as Bob's qubits 4 and 7, Eve is able to obtain full information about the key between the two legitimate communication partners without being noticed.

As a reaction Cabello published an addendum [14] to his protocol. There he described a solution to the problem, i.e., a way to secure the protocol in [13] against the ZLG attack. Cabello suggested to use the Hadamard operation H, which alters the Bell states as in the following way:

$$\begin{aligned} H|\Phi^{\pm}\rangle &= \frac{1}{\sqrt{2}}\Big(|\Phi^{\mp}\rangle \pm |\Psi^{\pm}\rangle\Big) = |\omega^{\pm}\rangle \\ H|\Psi^{\pm}\rangle &= \frac{1}{\sqrt{2}}\Big(|\Psi^{\mp}\rangle \pm |\Phi^{\pm}\rangle\Big) = |\chi^{\pm}\rangle . \end{aligned} \tag{5.38}$$

In detail, Alice and Bob exchange qubits 2 and 6 as in the original protocol but they perform their Bell state measurements afterward (cf. (2) and (3) in Fig. 5.6). Additionally, Alice decides randomly whether or not to perform a Hadamard operation on qubit 3 in her possession, which alters the state accordingly to

$$\begin{aligned} |\Psi^-\rangle_{12}|\chi^+\rangle_{35} = &\frac{1}{2}\Big(|\Phi^-\rangle_{13}|\omega^+\rangle_{25} - |\Phi^+\rangle_{13}|\omega^-\rangle_{25} \\ &+ |\Psi^-\rangle_{13}|\chi^+\rangle_{25} - |\Psi^+\rangle_{13}|\chi^-\rangle_{25}\Big). \end{aligned} \tag{5.39}$$

Alice announces her choice together with the result of her measurement on qubits 5 and 6. If Alice does not use the Hadamard operation, both parties follow the original protocol. Otherwise, Bob also performs a Hadamard operation on qubit 4 to undo the effects of the operation (cf. (4) in Fig. 5.6). Then he performs the Bell state measurement on qubits 2 and 4. Due to application of the Hadamard operation, Eve is not able to find a correct Pauli operation, when performing the ZLG attack, and thus Alice's and Bob's measurement results will not be correlated. If we take a look at the average collision probability for the ZLG attack, we get $\langle P_c \rangle = 0.75$, which is equal to the full I&R attack (cf. Eq. 5.27 in Sect. 5.2.1.3). Eve will be detected with a probability of $1 - (0.75)^n$ for n compared bits during the sifting phase, which can be made arbitrarily close to 1. The average information Eve learns about the intercepted qubits is the same as in the entanglement attack using the CNOT operation as described in Sect. 5.2.2.1. In case Alice does not apply the Hadamard operation, Eve is still able to obtain full information while staying undetected. Whenever Alice applies the Hadamard operation, Bob will not

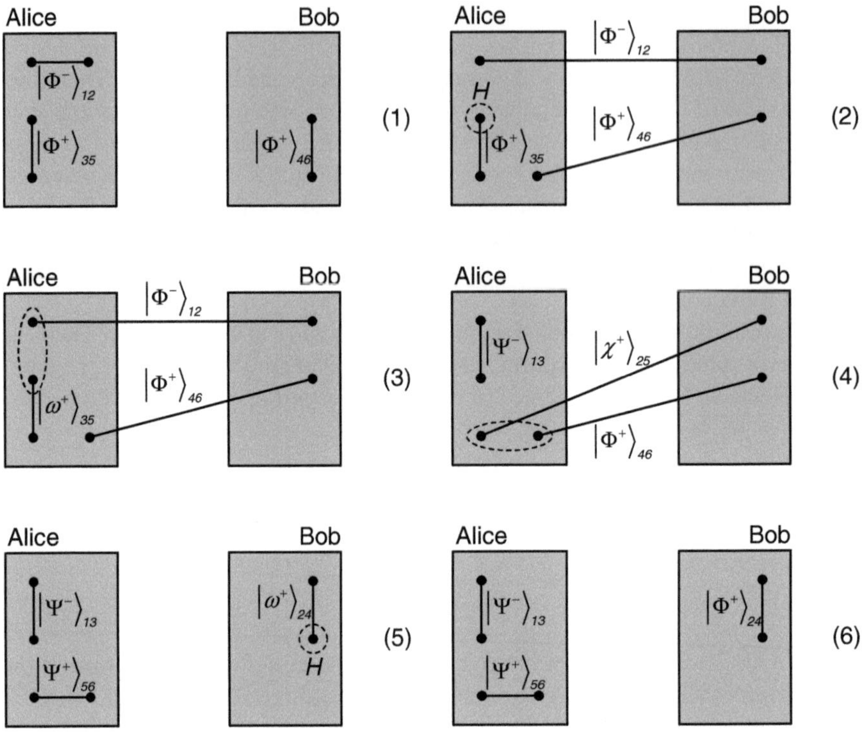

Fig. 5.6 Illustration of the revised protocol presented by Cabello [14]

get a correlated result from his measurement in 50% of the times. This gives Eve an information of $1 - H(S|M) = 0.75$ for each bit, but she will be detected easily due to the high and unbalanced occurrence of errors.

In the same year, Adan Cabello presented another protocol [12] for quantum key distribution and quantum secret sharing which is also open to a similar kind of attack. In this protocol three parties are involved which are able to distribute a key among them or share a secret between two of them. Therefore, each party is in possession of an entangled pair, i.e., $|\Phi^+\rangle_{12}$, $|\Phi^+\rangle_{4C}$, $|\Phi^+\rangle_{5D}$, and, additionally, Alice has a GHZ state $|\vartheta\rangle_{3AB}$ as given in Eq. 5.34 at her side (cf. (1) in Fig. 5.7). She keeps qubit 3 of the GHZ state and sends the other two qubits to Bob and Charly, respectively. Then, Alice performs a Bell state measurement on qubits 2 and 3, Bob performs his measurement on qubits 4 and A, and Charly performs his measurement on qubits 5 and B (cf. (2) in Fig. 5.7). As a consequence qubits 1, C, and D are now in a GHZ state due to entanglement swapping. It is assumed that Alice, Bob, and Charly obtain $|\Phi^+\rangle$ from their respective measurements which leaves qubits 1, C, and D still in the GHZ state $|\vartheta\rangle_{1CD}$, as presented in Fig. 5.7.

Bob and Charly send their remaining qubits C and D back to Alice, who performs a GHZ state measurement and publicly announces the outcome. Based on

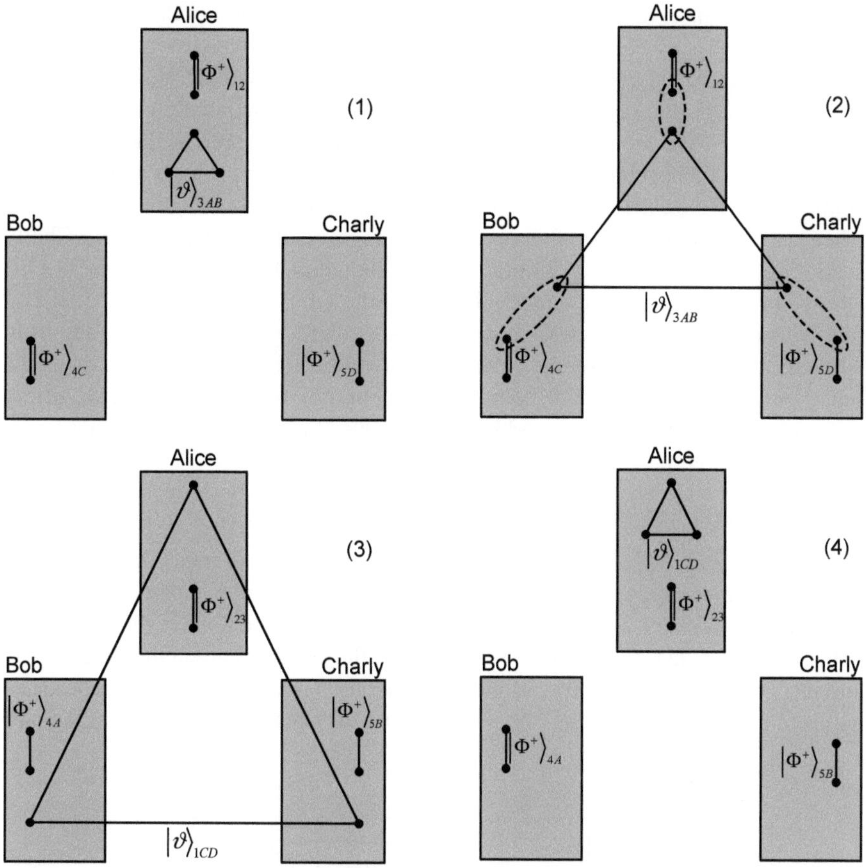

Fig. 5.7 Illustration of the protocol in [12]

this public result and the results of their own measurements all three parties can agree on a single classical bit which is later used for the secret key.

It has been shown by Lee et al. [21] that this protocol is also open to the ZLG attack strategy. In detail, Eve prepares two entangled pairs in the state $|\Phi^+\rangle_{PQ}$ and $|\Phi^+\rangle_{RS}$ and intercepts qubits A and B coming from Alice. She keeps the qubits P and R and forwards qubit Q and qubit S to Bob and Charly, respectively. Both parties perform their measurement as described in the protocol and they return the qubits C and D. Eve intercepts also these qubits and performs a Bell measurement on the pairs P, C and R, D. According to the measurement results, Eve is able to select a Pauli operator and apply it on the qubits A and B that she intercepted from Alice to preserve the correlation. Since they are still in a GHZ state together with one qubit from Alice, these operations alter the overall state in a way such that it corresponds to Alice's, Bob's, and Charly's measurement results. In the end Eve returns the two qubits to Alice, who performs a GHZ state measurement on them as described in the protocol. The three legitimate communication parties will not detect Eve because, due to her Pauli operations, she does not introduce any error in the protocol.

In their paper [21] Lee et al. also presented a method to secure Cabello's protocol against the ZLG attack. In this case Bob and Charly use the quantum Fourier transformation (QFT) to secure the qubits in transit. After they received the qubits from Alice each of them returns one qubit of the entangled pair to Alice and randomly applies the QFT on the other one. Then they publicly announce their decision and Alice performs the inverse QFT on the qubits she received from Bob and Charly according to their decision. Similar to the addendum to Cabello's protocol [14] Eve is not able to overcome the application of the QFT by Bob and Charly. If Eve follows the attack strategy described in [21] she intercepts the qubits coming from Alice, Bob, and Charly but she cannot find a Pauli operation to correct the GHZ state. Thus, the three legitimate communication parties detect Eve during the sifting phase of the protocol due to the additional amount of error introduced by her.

5.2.2.3 General Entanglement Swapping Attack

Another protocol, which was presented by Li et al. [22], also uses entangled states and entanglement swapping to distribute a secret key between two parties. Instead of three entangled states as in [13] just two entangled pairs are used in this protocol – both at Alice's side (cf. (1) in Fig. 5.8). Initially, the two pairs are in the state $|\Phi^+\rangle_{12}$ and $|\Phi^+\rangle_{34}$ and Alice performs some Pauli operation σ on qubit 1 before she sends qubits 2 and 4 to Bob. Then, she performs a Bell state measurement on the two

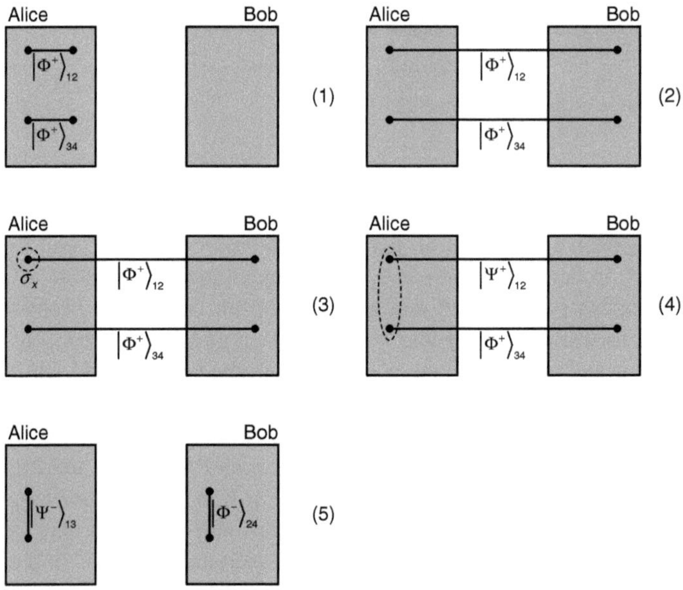

Fig. 5.8 Illustration of the protocol presented in [22]

remaining qubits in her possession (cf. (3) and (4) in Fig. 5.8. Due to entanglement swapping she knows exactly the state qubits 2 and 4 are in (cf. Eq. 5.40):

$$\begin{aligned}\sigma|\Phi^+\rangle_{12}|\Phi^+\rangle_{34} = \frac{1}{2}\Big(&|\Phi^+\rangle_{13}\sigma|\Phi^+\rangle_{24} + |\Phi^-\rangle_{13}\sigma|\Phi^-\rangle_{24}\\ &+ |\Psi^+\rangle_{13}\sigma|\Psi^+\rangle_{24} + |\Psi^-\rangle_{13}\sigma|\Psi^-\rangle_{24}\Big)\end{aligned}. \tag{5.40}$$

Further, Alice is able to compute the state Bob's qubits would be in if she had not performed her Pauli operation. This piece of information is called Alice's *imaginary result* by Li et al. in [22]. When Bob performs a Bell state measurement on qubits 2 and 4 he is able to guess Alice's imaginary result based on his own result. Alice then publicly reveals her measurement result such that Bob can calculate the exact operation σ she has applied before. Alice and Bob use their secret results to generate a classical secret key.

In [35] it has been shown that this protocol is insecure against an attack strategy based on entanglement swapping. The attack applied in this case is a kind of generalization of the ZLG attack [38] since it is based on an entangled six-qubit state $|\delta\rangle$ (cf. Eq. 5.41) instead of just a single pair and it is also applicable on the original Cabello protocols presented above [13, 12]

$$\begin{aligned}|\delta\rangle_{PQRSTU} = \frac{1}{2\sqrt{2}}\Big(&|HHHHHH\rangle_{PQRSTU} + |HHVVHV\rangle_{PQRSTU}\\ &+ |HVHVVV\rangle_{PQRSTU} + |HVVHVH\rangle_{PQRSTU}\\ &+ |VHHVVH\rangle_{PQRSTU} + |VHVHVV\rangle_{PQRSTU}\\ &+ |VVHHHV\rangle_{PQRSTU} + |VVVVHH\rangle_{PQRSTU}\Big).\end{aligned} \tag{5.41}$$

The main idea of this attack is that the six-qubit state preserves the correlations between the measurement results coming from the entanglement swapping. This fact allows Eve to stay undetected. Furthermore, Eve is able to keep two qubits of the six-qubit state which are in the same state as Bob's qubits after the protocol has finished. Explicitly, that means

$$\begin{aligned}|\delta\rangle_{PQRSTU} = \frac{1}{2}\Big(&|\Phi^+\rangle_{PR}\otimes|\Phi^+\rangle_{QS}\otimes|\Phi^+\rangle_{TU}\\ &+ |\Phi^-\rangle_{PR}\otimes|\Phi^-\rangle_{QS}\otimes|\Phi^-\rangle_{TU}\\ &+ |\Psi^+\rangle_{PR}\otimes|\Psi^+\rangle_{QS}\otimes|\Psi^+\rangle_{TU}\\ &+ |\Psi^-\rangle_{PR}\otimes|\Psi^-\rangle_{QS}\otimes|\Psi^-\rangle_{TU}\Big).\end{aligned} \tag{5.42}$$

This gives Eve the same information Bob has at that time. Further, Eve listens to the public communication between Alice and Bob and thus obtains full information about the secret key. Eve will not be detected during the error correction phase because the correlation between Alice's and Bob's results is preserved by the state $|\delta\rangle$.

Regarding the protocol presented in [22] Eve's attack strategy is the following: Eve intercepts qubit 2 coming from Alice and qubit 3 coming from Bob (the initial settings of [22] have been altered in [35] due to a rather simple loophole in the original protocol). Eve performs a Bell state measurement on qubits 2 and P and qubits 3 and S and in this way entangles herself with the two communication parties (cf. (2) and (3) in Fig. 5.9). Eve then sends qubits Q to Bob and R to Alice, such that all three parties are in possession of two qubits of the six-qubit state $|\delta\rangle$. When Bob performs a Bell state measurement on his two qubits 4 and Q the remaining four qubits of $|\delta\rangle$ collapse into two entangled pairs (cf. (6) and (7) in Fig. 5.9). The state of these pairs is completely determined by Bob's measurement result. As pointed out in Eq. 5.42 the two qubits at Eve's side, T and U, are in the same state as Bob's qubits 4 and Q, and the two qubits at Alice's side, 1 and R, are perfectly correlated to Bob's result as it would be expected (cf. (7) in Fig. 5.9).

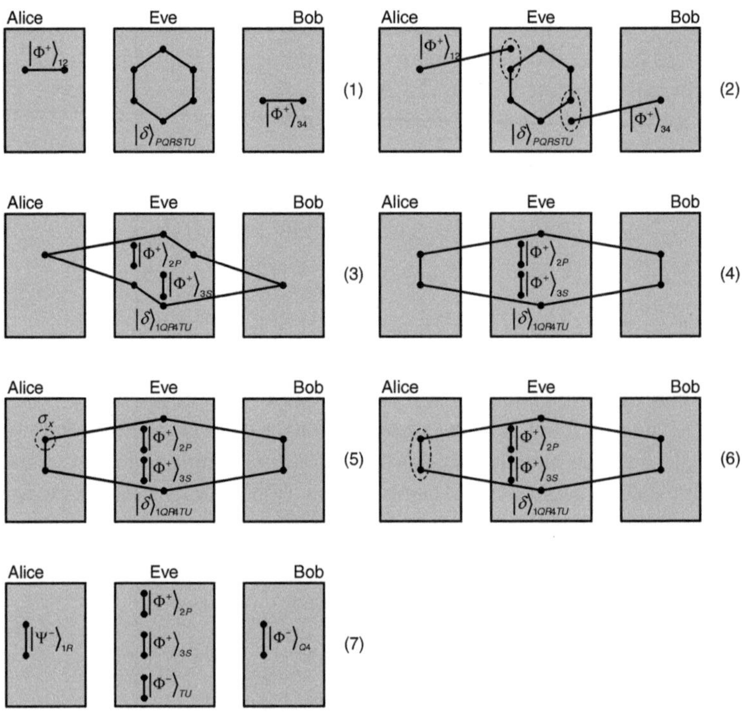

Fig. 5.9 Illustration of the attack strategy presented in [35]

It is also described in [35] how the protocol can be secured against such an attack strategy: it is the same way Cabello secured his protocol in [14] – by using a Hadamard operation. In the beginning of the modified protocol Alice randomly chooses whether to perform the Hadamard operation onto qubit 1 or not. This alters the initial state of the original protocol to

$$|\omega^+\rangle_{12}|\Phi^+\rangle_{34} = \frac{1}{2}\Big(|\Phi^+\rangle_{13}|\omega^+\rangle_{24} + |\Phi^-\rangle_{13}|\omega^-\rangle_{24} + |\Psi^+\rangle_{13}|\chi^+\rangle_{24} + |\Psi^-\rangle_{13}|\chi^-\rangle_{24}\Big), \tag{5.43}$$

where $|\omega^\pm\rangle$ and $|\chi^\pm\rangle$ come from Eq. 5.38. Both parties then follow the protocol as described above until Alice receives the qubit coming from Bob. Then she publicly announces her choice on the Hadamard operation and Bob applies the Hadamard operation on qubit 2, if necessary, to undo it. Both perform their Bell state measurements and their results are correlated as it would be expected and they are able to extract a classical secret key as described in [22]. Similar to the scenarios described above, Eve will not be able to preserve the correlation between Alice's and Bob's result and thus will introduce an error every time Alice performs the Hadamard operation. In detail, Eve's average collision probability is again $\langle P_c\rangle = 0.75$ as for the full I&R attack (cf. Eq. 5.27 in Sect. 5.2.1.3).

Thus, Alice and Bob can make the probability of detecting Eve arbitrarily close to 1, i.e., $1 - (0.75)^n$. As already pointed out in Sect. 5.2.2.1 and 5.2.2.2, Eve is still able to perfectly eavesdrop the secret bit whenever Alice does not perform the Hadamard operation. This gives Eve the average information $1 - H(S|M) = 0.75$, which is higher than in the full I&R attack, but she will be detected easily because of the 50% additional error rate, which is introduced when Alice uses the Hadamard operation.

As it is further described in [35], Eve can find a state $|\delta'\rangle$ which can compensate the application of the Hadamard operation. But in this case Eve will introduce the same amount of error every time Alice's does not use the Hadamard operation. Thus, Eve will gain nothing by using $|\delta'\rangle$ instead of $|\delta\rangle$.

5.3 Individual Attacks in an Realistic Environment

The protocols and attacks described above are settled in an ideal environment. The photon sources emit single-photon signals only and the detectors are 100% efficient. But using today's technology such a setting is impossible to achieve as we have discussed in Chap. 6. Detectors are highly sensitive and often detect a signal even if none was sent (these events are called *dark counts*). This has to be considered when performing the error correction and privacy amplification.

Further, there are no single-photon sources but a normal signal pulse often contains a large number of photons. To solve this problem, weak coherent pulses (WCP) are used in actual quantum cryptographic devices (cf., for example, [30]), which are described as

$$|\alpha\rangle = e^{\frac{-|\alpha|^2}{2}} \sum_{n=0}^{\infty} \frac{\alpha^n}{\sqrt{n!}}|n\rangle, \tag{5.44}$$

which is a superposition of Fock states (states with 0, ..., n photons). Such pulses have a rather low mean photon number μ, such that the probability to find more than one photon in a pulse follows a Poissonian distribution [16]

$$P(n, \mu) = \frac{\mu^n}{n!} e^{-\mu}. \tag{5.45}$$

The mean photon number μ cannot be made arbitrarily low, because this will decrease the efficiency of the protocol.

In the following some attack strategies will be presented which make use of such loopholes given by the physical limitations of QKD protocols.

5.3.1 PNS Attack

The *photon number splitting attack* (PNS) was first introduced by Huttner et al. [18] and later discussed by Brassard et al. [10] and Lütkenhaus [25] and is the most powerful individual attack. It is applied on realistic photon sources emitting weak coherent pulses which generate single photons only with a certain probability (see the paragraph above). With a small probability multi-photon pulses are emitted containing two or more photons having the same polarization. The strategy for Eve is to intercept these pulses coming from Alice, take one photon of the multi-photon pulse and send the remaining photon(s) along to Bob. Eve waits until Alice and Bob publicly compare their measurement bases and then measures the intercepted photon in the correct basis.

In detail, the PNS attack is a little more complex. According to Eq. 5.45, the probability that Alice's source emits a vacuum pulse (containing zero photons) is very high and the probability of a single-photon pulse is around 10%. Hence, the probability of a multi-photon pulse is very low (around 5% [16]). Because of this Eve cannot split a photon off each pulse but she has to check for the multi-photon pulses. Therefore, she performs a non-demolition measurement to collapse the pulse into a state containing a fixed number of photons. This is accomplished by a projection onto Fock spaces. If Eve intercepted a multi-photon pulse, she applies an operator A_N, which destructs one photon of the pulse and creates an appropriate auxiliary state (cf. Eq. 5.46)

$$\begin{aligned} A_N|N,0\rangle_+|\alpha\rangle &= |N-1,0\rangle_+|\varphi_1\rangle \\ A_N|0,N\rangle_+|\alpha\rangle &= |0,N-1\rangle_+|\varphi_2\rangle \\ A_N|N,0\rangle_\times|\alpha\rangle &= |N-1,0\rangle_\times|\Psi_1\rangle \\ A_N|0,N\rangle_\times|\alpha\rangle &= |0,N-1\rangle_\times|\Psi_2\rangle \end{aligned} . \tag{5.46}$$

From her measurement on the auxiliary system together with the information about Alice's basis choice, Eve is able to determine the correct value of the secret bit. Therefore, $\langle\varphi_1|\varphi_2\rangle$ and $\langle\Psi_1|\Psi_2\rangle$ have to be zero such that they can be distinguished by Eve. As pointed out in [10] such an operator can be described by the Jaynes–Cummings model.

Using the operator A_N Eve is able to obtain full information from multi-photon signals generated by Alice's source. But, as we already pointed out, the probability that a multi-photon signal is emitted is rather small. Only if the probability that Bob detects a signal is smaller than the probability of a multi-photon signal the attack becomes a severe problem. In this case Eve suppresses all dark counts in Bob's module and the efficiency of his detectors is increased to 100%. Further, Eve replaces the quantum channel with a perfect channel, i.e., there are no losses due to the channel. It is a rather paranoid assumption to give Eve the power to do all these things, since they affect Bob's hardware directly. But, to be secure, all possible scenarios have to be considered. For each signal coming from Alice, Eve acts in the following way: all signals with zero photons are ignored, since dark counts have been suppressed. All multi-photon signals are attacked using the PNS strategy. This gives Eve full information about the corresponding bit of the secret key. A fraction of the single-photon signals is suppressed and the other single-photon signals are attacked using the I&R strategy (cf. Sect. 5.2.1). Eve chooses the amount of discarded signals such that they are consistent with Bob's total detection probability. With a perfect quantum channel and perfect detectors all errors in this scenario are introduced by Eve's I&R attack (the PNS attack introduces no error). Bob is not able to distinguish these errors from the ones he expects due to dark counts and the lossy channel. In this case the whole communication becomes insecure.

In [26] it has been shown that also the Poisson photon number distribution can be preserved using the PNS attack, which makes it undetectable as long as a publicly known signal intensity is used. Therefore, the decoy states method [19, 23, 37] uses different intensities to make a detection of the PNS attack possible (cf. "Sect. 4.2.3"). Another way to secure BB84-like protocols against the PNS attack was presented in [33]. Scarani et al. suggested an alternative sifting procedure such that Alice does not give away her measurement basis. Instead, she announces one of four pairs of non-orthogonal states. This leaves Bob with an inconclusive or ambiguous result and he will have to discard his result for 75% of all signals. Although the efficiency of this protocol is much lower than for standard BB84 protocols (where about 50% of the signals are discarded), it gives not enough information to an eavesdropper such that the PNS attack can be applied successfully.

5.3.2 Trojan Horse Attack

Another attack strategy on realistic setups of QKD systems is the *Trojan Horse attack* or *light injection attack*. It has been introduced first in [32, 6] and was discussed in more detail in [36] later on. The main idea of this attack strategy is not to interact with the photons in transit between Alice and Bob but to probe the devices in Alice's and Bob's laboratory by sending some light into them and collecting the reflected signal. In this way Eve is able to obtain information about the detectors and further on which classical bit Bob measured. In detail, Eve is in possession of a laser and a detection scheme. She sends out light pulses toward Alice's or Bob's setup, which are reflected and enter the detection scheme when returning to Eve. In [36] it is assumed that Eve uses homodyne detection for the reflected pulse and thus needs a reference pulse. This reference pulse is delayed in an arm of the optical fiber and enters the detection system together with the reflected pulse.

Eve can use the information of the reflected signal to detect which basis Alice's used for the preparation of the photon. The detection of the correct basis is based on a phase modulation occurring due to the different ways the reflected and reference beams go through [36]. If Eve is able to do this before Alice's photon reaches Bob, she can perform a simple I&R attack (cf. Sect. 5.2.1), i.e., intercept the photon in transit, measure it in the correct basis, and send it on to Bob. This will give her full information on the secret bit string.

A countermeasure against this kind of attack strategy is implemented in the plug and play systems (cf. Chap. 6) where the intensity of incoming light is monitored [32, 6]. The idea is that Bob sends a rather intense beam of light to Alice which is used for synchronization with a special timing detector at Alice's setup. This detector notifies the legitimate communication parties when the power of an incoming signal extends some predefined level. For protocols where light just goes one way (e.g., out of Alice's lab into Bob's lab) a strategy for preventing the attack is to add components in Alice's and Bob's laboratory to block Eve's injected pulse. This means, for example, that the laser pulses have to pass through an optical isolator and a band-pass filter [36] when leaving Alice's setup. The isolator reduces the signals coming into Alice's laboratory to make a light injection attack impossible.

5.3.3 Faked States Attack

The *faked states* attack is a kind of I&R attack strategy but Eve does not try to recreate the intercepted state. Instead, Eve manages to send a signal to Bob which he can only detect in a way totally controlled by Eve. This attack was first introduced in [28] and later extended in [27, 29]. In detail, Eve intercepts the signals coming from Alice using an apparatus similar to Bob's. Further, she forwards a state to Bob which can only be detected by him if he chooses the same basis as Eve. She can achieve this by exploiting the *full detector efficiency mismatch* [27]. This is a phenomenon where the signal coming into the detector has a time shift such that it is outside the detector's sensitivity curve. Therefore, only one detector can fire and the other one is blinded out. In this way Eve can control the bit value Bob will obtain from his measurement. The second goal of the faked states attack is to eliminate the case where Bob performs a measurement in a basis incompatible to Eve's basis, thus detecting an error. Eve can achieve that by adding a relative phase to the signal such that the whole signal is deflected to the blinded detector and is lost.

For the BB84 protocol [2] the faked states attack works as follows: Eve performs an I&R attack and obtains some result from her measurement. Then she sends a signal pulse to Bob which has the opposite bit value in the opposite basis compared to what she has detected. Eve also sets the time shift of the signal such that the detector for the opposite bit value compared to what she has detected is blinded out. Thus, if Bob tries to detect the signal in a different basis than Eve, he would not detect anything. Otherwise, if Bob chooses the same basis as Eve, he will either detect the same bit as Eve or nothing at all. Therefore, every time Eve measured Alice's state in the wrong basis, also Bob will measure it in the wrong basis and the results will be discarded. If Eve has chosen the right basis, also Bob measured in the right basis and Eve has full information about this bit of the secret key.

As explained in [27, 29] it has to be stressed that Bob's detection efficiency is reduced by the faked states attack since all signals where Bob measured in a different basis compared to Eve and half of the signals where Bob measured in the same basis are suppressed. Eve can overcome this rather easily using faked states with a proportionally increased brightness. If Eve is not able to blind one detector completely, she can only obtain partial information about the key but, nevertheless, stays undetected [31].

Possible countermeasures to prevent the attack are, for example, to actively monitor the timing of incoming pulses at Bob's side [27]. This can be achieved through a random shifting of Bob's time window or with additional detectors. Alternatively, Bob can test the characteristics of his detectors over a variety of input signals to especially check all features of the sensitivity curve. Another countermeasure for Bob is to introduce random jitter into the detector synchronization to smear the curves and lower the mismatch.

5.3.4 Time-Shift Attack

An alternative version of the faked states attack is the *time-shift attack strategy* [31]. The time-shift attack also exploits the detector efficiency mismatch, but, contrary to the faked state attack [28, 27, 29], it is feasible with today's technology, as it has been shown in [39]. The main difference is that Eve does not measure the state in transit between Alice and Bob but randomly shifts the time of the signal such that it arrives outside of Bob's detector's sensitivity curve. Due to her choice of the time delay, Eve is able to infer the exact result of Bob's measurement. As pointed out in Sect. 5.3.3, if Eve is able to completely blind a detector by her time shift, she is able to obtain full information about Bob's measurement result. Otherwise, Eve will obtain only partial information about the secret key. In both cases, Eve never introduces any error, since she does not measure or otherwise interact with Alice's state in transit.

One difference to the faked states attack is that Eve has to deal with the increased loss at Bob's side in another way. Regarding the faked states attack Eve uses a brighter laser pulse to overcome the losses, as described in Sect. 5.3.3. With respect to the time-shift attack Eve has to replace the quantum channel by a low-loss version to compensate Bob's additional losses.

The countermeasures described in Sect. 5.3.3 will also work here to prevent an application of the time-shift attack. Additionally, phase shift settings can be applied to Bob's phase modulator and the detection rate and the channel loss can be checked to secure a protocol against the time-shift attack [31].

References

1. Bennett, C.H.: Quantum cryptography using any two nonorthogonal states. Phys. Rev. Lett. **68**(21), 3121–3124 (1992)
2. Bennett, C.H., Brassard, G.: Public key distribution and coin tossing. In: Proceedings of the IEEE International Conference on Computers, Systems, and Signal Processing, pp. 175–179. IEEE Press, New York (1984)
3. Bennett, C.H., Brassard, G., Breidbart, S., Wiesner, S.: Quantum Cryptography, or Unforgeable Subway Tokens. Advances in Cryptology: Proceedings of the Crypto '82, pp. 267–275 (1982)
4. Bennett, C.H., Brassard, G., Crepeau, C., Jozsa, R., Peres, A., Wootters, W.K.: Teleporting an unknown quantum state via dual classical and EPR channels. Phys. Rev. Lett. **70**(13), 1895–1899 (1993)
5. Bennett, C.H., Brassard, G., Popescu, S., Schumacher, B., Smolin, J., Wootters, W.K.: Purification of noisy entanglement and faithful teleportation via noisy channels. Phys. Rev. Lett. **76**(5), 722–725 (1996)
6. Bethune, D.S., Risk, W.P.: An autocompensating fiber-optic quantum cryptography system based on polarization splitting of light. IEEE J. Quantum Electron. **36**(3), 340–347 (2000)
7. Biham, E., Boyer, M., Brassard, G., van de Graf, J., Mor, T.: Security of quantum key distribution against all collective attacks. Algorithmica **34**(4), 372–388 (2002)
8. Biham, E., Mor, T.: Security of quantum cryptography against collective attacks. Phys. Rev. Lett. **78**(11), 2256–2259 (1997)
9. Bose, S., Vedral, V., Knight, P.L.: Multiparticle generalization of entanglement swapping. Phys. Rev. A **57**(2), 822–829 (1998)
10. Brassard, G., Lütkenhaus, N., Mor, T., Sanders, B.C.: Limitations on practical quantum cryptography. Phys. Rev. Lett. **85**(6), 1330–1333 (2000)
11. Bruss, D.: Optimal eavesdropping in quantum cryptography with six states. Phys. Rev. Lett. **81**(14), 3018–3021 (1998)
12. Cabello, A.: Multiparty key distribution and secret sharing based on entanglement swapping. quant-ph/0009025 v1 (2000)
13. Cabello, A.: Quantum key distribution without alternative measurements. Phys. Rev. A **61**(5), 052,312 (2000)
14. Cabello, A.: Reply to "comment on quantum key distribution without alternative measurements". Phys. Rev. A **63**(3), 036,302 (2001)
15. Ekert, A.: Quantum cryptography based on Bell's theorem. Phys. Rev. Lett. **67**(6), 661–663 (1991)
16. Gisin, N., Ribordy, G., Tittel, W., Zbinden, H.: Quantum cryptography. Rev. Mod. Phys. **74**(1), 145 (2002)
17. Greenberger, D., Horne, M.A., Zeilinger, A.: Going beyond Bell's Theorem. In: M. Kafatos (ed.) Bell's Theorem, Quantum Theory and Conceptions of the Universe, pp. 69–72. Kluwer, Dordrecht (1989)
18. Huttner, B., Imoto, N., Gisin, N., Mor, T.: Quantum cryptography with coherent states. Phys. Rev. A **51**(3), 1863–1869 (1995)
19. Hwang, W.Y.: Quantum key distribution with high loss: Toward global secure communication. Phys. Rev. Lett. **91**(5), 057,901 (2003)

20. Kim, Y.H., Kulik, S., Shih, Y.: Quantum teleportation of a polarization state with complete bell state measurement. Phys. Rev. Lett. **86**(7), 1370–1373 (2001)
21. Lee, J., Lee, S., Kim, J., Oh, S.D.: Entanglement swapping secures multiparty quantum communication. Phys. Rev. A **70**(3), 032,305 (2004)
22. Li, C., Wang, Z., Wu, C.F., Song, H.S., Zhou, L.: Certain quantum key distribution achieved by using Bell states. Int. J. Quantum Inf. **4**(6), 899–906 (2006)
23. Lo, H.K., Ma, X., Chen, K.: Decoy state quantum key distribution. Phys. Rev. Lett. **94**(23), 230,504 (2005)
24. Lütkenhaus, N.: Security against eavesdropping attacks in quantum cryptography. Phys. Rev. A **54**(1), 97–111 (1996)
25. Lütkenhaus, N.: Security against individual attacks for realistic quantum key distribution. Phys. Rev. A **61**(5), 052,304 (2000)
26. Lütkenhaus, N., Jahma, M.: Quantum key distribution with realistic states: Photon-number statistics in the photon-number splitting attack. New J. Phys. **4**, 44.1–44.9 (2002)
27. Makarov, V., Anisimov, A., Skaar, J.: Effects of detector efficiency mismatch on security of quantum cryptosystems. Phys. Rev. A **74**(2), 022,313 (2006)
28. Makarov, V., Hjelme, D.R.: Faked states attack on quantum cryptosystems. J. Mod. Opt. **52**(5), 691–705 (2005)
29. Makarov, V., Skaar, J.: Faked states attack using detector efficiency mismatch on SARG04, phase-time, DPSK and Ekert protcols. Quant. Inf. Comp. **8**(6&7), 622–635 (2008)
30. Poppe, A., Peev, M., Maurhart, O.: Outline of the SECOQC quantum-key-distribution network in Vienna. Int. J. of Quant. Inf. **6**(2), 209–218 (2008)
31. Qi, B., Fung, C.H.F., Lo, H.K., Ma, X.: Time-shift attack in practical quantum cryptosystems. Quant. Inf. Comp. **7**(1&2), 73–82 (2007)
32. Ribordy, G., Gautier, J.D., Gisin, N., Guinnard, O., Zbinden, H.: Fast and user-friendly quantum key distribution. J. Mod. Optics **47**(2&3), 517–531 (2000)
33. Scarani, V., Acin, A., Ribordy, G., Gisin, N.: Quantum cryptography protocols robust against photon number splitting attacks for weak laser pulses implementations. Phy. Rev. Lett. **92**(5), 057,901 (2004)
34. Scarani, V., Bechmann-Pasquinucci, H., Cerf, N.J., Dusek, M., Lütkenhaus, N., Peev, M.: The Security of Practical Quantum Key Distribution. quant-ph/0802.4155 v2 (2008)
35. Schauer, S., Suda, M.: A novel attack strategy on entanglement swapping QKD protocols. Int. J. Quant. Inf. **6**(4), 841–858 (2008)
36. Vakahitov, A., Makarov, V., Hjelme, D.R.: Large pulse attack as a method of conventional optical eavesdropping in quantum cryptography. J. Mod. Opt. **48**(13), 2023–2038 (2001)
37. Wang, X.B.: Beating the photon-number-splitting attack in practical quantum cryptography. Phys. Rev. Lett. **94**(23), 230,503 (2005)
38. Zhang, Y.S., Li, C.F., Guo, G.C.: Comment on "quantum key distribution without alternative measurements". Phys. Rev. A **63**(3), 036,301 (2001)
39. Zhao, Y., Fung, C.H.F., Qi, B., Chen, C., Lo, H.K.: quantum hacking: Experimental demonstration of time-shift attack against practical quantum key distribution systems. Phys. Rev. A **78**(4), 042,333 (2008)
40. Zukowski, M., A., Z., Horne, M.A., Ekert, A.K.: "Event-Ready-Detectors" Bell state measurement via entanglement swapping. Phys. Rev. Lett. **71**(26), 4287–4290 (1993)

Chapter 6
QKD Systems

M. Suda

6.1 Introduction

This chapter summarizes information about seven QKD systems that have been developed for SECOQC [58, 1, 26] and which are candidates for further integration to build up a prototype QKD network. In Chap. 9 the quantum-cryptographic network of SECOQC is described in detail and Chap. 7 presents a statistical analysis of the network in a real-life environment. The mentioned seven systems are in Sect. 6.2.1 the plug and play system (PP), in Sect. 6.2.2 the phase-coding QKD system or one-way weak coherent pulse QKD system (Toshiba), in Sect. 6.2.3 the time-coding QKD system or coherent one-way system (COW), in Sect. 6.2.4 the continuous variables system or QKD with coherent states (CV), in Sect. 6.2.5 the entanglement-based QKD system (EB), in Sect. 6.2.6 the free-space QKD system (FS), and finally in Sect. 6.2.7 the low-cost QKD system (LC).

In each section the basic ideas of each system and a more detailed physical description are presented. Some technical details are given which are supplied by the experimental groups themselves. Overviews of the underlying physical principles of the systems and relating quantum protocols can be found, e.g., in [27] and [18]. Since security is the core business of a QKD network we refer to [71] which gives an overview of the security of practical QKD.

The theoretical modeling of each system is described and the status of the security proofs for various scenarios is mentioned. Implications of the known security results are demonstrated. For some systems the quantum bit error rate and the secure key rate as a function of distance between the communicating partners Alice and Bob are discussed.

M. Suda (✉)
Safety & Security Department, Quantum Technologies, AIT Austrian Institute of Technology GmbH, Donau-City-Straße 1/1220 Vienna, Austria,
martin.suda@ait.ac.at; www.ait.ac.at

Suda, M.: *QKD Systems*. Lect. Notes Phys. **797**, 97–121 (2010)
DOI 10.1007/978-3-642-04831-9_6 © Springer-Verlag Berlin Heidelberg 2010

6.2 QKD Systems

In the following, the technical and physical aspects of the 7 QKD systems developed within SECOQC are described. Their physical modes of operation are explained and information is given about the capability of the systems and about security aspects including the quantum bit error rate and the achievable secure bit rate. The applied quantum protocol is discussed. Appropriate literature references are enclosed.

6.2.1 Plug and Play (PP)

Since the introduction of the BB84 protocol by Bennett and Brassard [6] and the first realization in 1992 [5], many experiments have been undertaken relating QKD (see, e.g., [27]). In this section we describe a fiber-optic QKD prototype which works as an auto-compensating plug and play system [76, 59]. The device is a long-distance (67 km) QKD system employing optical fibers and works at telecom wavelengths (1550 nm) using standard telecom components. The qubits are encoded in the relative phase between two subsequent pulses and analyzed by an unbalanced interferometer with active phase modulation. The auto-compensation is related to polarization rotations in the fiber.

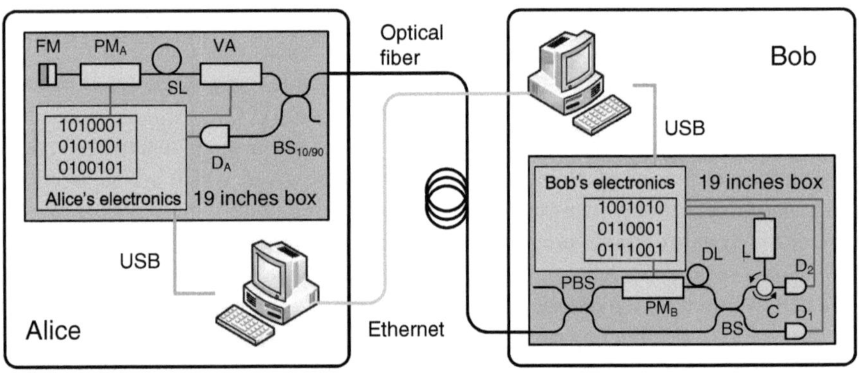

Fig. 6.1 Sketch of plug and play system; L: laser, C: circulator, BS: 50/50 beam splitter, DL: delay line, PM_B: Bob's phase modulator, PBS: polarizing beam splitter, $BS_{10/90}$: 10/90 beam splitter, VA: variable attenuator, SL: storage line, PM_A: Alice's phase modulator, FM: Faraday mirror, D: detector; see text for details

The plug and play system is sketched in Fig. 6.1. Strong linearly polarized pulses of photons are created by a laser L on Bob's side. The frequency of the pulses is 5MHz (period of 200 ns). The beam is separated into two parts at the 50/50 beam splitter BS. The long arm contains a delay line DL of a length of 10 m (50 ns $\doteq$ 20 MHz). The phase modulator PM_B is not used at that time. In the shorter arm the linear polarization is turned by 90° (not visible in Fig. 6.1). Both beams are recombined at the polarizing beam splitter PBS where they exit Bob's setup one after another (time delay of 50 ns) by the same port because of the orthogonal polarization states of the two pulses. Thereby the first pulse passed the short and the second pulse the long arm of the interferometer.

The pulses travel down to Alice where they, passing a $BS_{10/90}$ (90% of the intensity is registered in the detector D_A), are attenuated (variable attenuator VA) and reflected on a Faraday mirror FM (here the polarization states are reversed) and are further attenuated by VA. The storage line SL will be discussed below. Moreover, Alice applies a phase of 0 or π and $\frac{\pi}{2}$ or $\frac{3\pi}{2}$ on the second pulse (thus implementing the BB84 protocol) with the phase modulator PM_A. At the output of Alice's setup the polarizations of the two pulses are again orthogonal to each other, but have been interchanged because of the FM. Thus, a compensation of all accumulated polarization rotations (Bob to Alice) can take place on the way back from Alice to Bob (auto-compensating system).

Arriving at Bob's interferometer – because of the changed polarization states – the first pulse now enters the long arm where Bob chooses the measurement basis by applying a 0 or a $\frac{\pi}{2}$ phase shift on its way back using PM_B. The second pulse takes the short path. Both pulses arrive at the same time at the BS where they interfere. Then they are detected either in D_1 or, after passing through the circulator C, in D_2.

On the way back from Alice to Bob the plug and play system is a usual QKD system using phase encoding between coherent pulses. The strong pulses sent from Bob to Alice do not contain the information about the qubit: the quantum information travels only one way, namely from Alice to Bob.

Since the pulses travel back and forth, backscattering light (elastic Rayleigh scattering) can considerably increase the noise. Therefore, Bob's laser sends trains of pulses. The length of these trains corresponds to the length of the storage line (SL) introduced for this purpose behind the VA at Alice's setup. Therefore, the backward propagating pulses do no longer cross the bright pulses in the fiber. For a length of the SL of approximately 20 km, a pulse train contains 480 pulses at a frequency of 5 MHz. The 10/90 BS directs most of the incoming light to a detector module D_A which is – amongst others – used to synchronize the 5 MHz clock of Bob's laser and which guarantees that PM_A is activated in time. This synchronized clock allows Alice to apply a phase shift exactly when the second pulse passes 50 ns after the first pulse. This second pulse contains phase information and must be attenuated below the one-photon-per-pulse level.

As a measure of security, the number of coincident clicks at both detectors D_1 and D_2 is registered which is important to limit beam-splitting attacks.

The raw key R_{raw} between Alice, the transmitter, and Bob, the receiver, is

$$R_{\mathrm{raw}} = q\,\nu\,\mu\,t_{AB}\,t_B\,\eta_B\,\eta_{SL}\,\eta_\tau, \tag{6.1}$$

where the following quantities have been used:
q, depends on the implementation ($= \frac{1}{2}$ for the BB84 protocol)
ν, repetition frequency
μ, average number of photons per pulse (≈ 0.1)
t_{AB}, transmission of the line Alice–Bob ($= 10^{-\alpha d/10}$, $\alpha = 0.2\,\mathrm{dB/km} =$ absorption,

d = distance)
t_B, Bob's internal transmission (≈ 0.6)
η_B, Bob's detection efficiency (≈ 0.1)
η_{SL}, factor because of length l_{SL} of the storage line ($= l_{SL}/l_{SL} + d$)
η_τ, factor because of dead time τ of the detector ($\lessapprox 1$).

The second-most important parameter is the quantum bit error rate $QBER = \frac{false\ counts}{total\ counts}$ (should be $< 10\%$):

$$QBER = QBER_{\text{opt}} + QBER_{\text{dark}} + QBER_{\text{after}} + QBER_{\text{stray}} \ . \qquad (6.2)$$

$QBER_{\text{opt}}$ ... probability for the photon to hit the wrong detector
$QBER_{\text{dark}}$ ... error rate because of dark counts [77]
$QBER_{\text{after}}$ ... is the probability to have an after pulse in the detector [77]
$QBER_{\text{stray}}$ ... the errors induced by stray light (Rayleigh backscattering)

Error correction and privacy amplification lead to the following formula of the final key rate R_{fin} [27, 20, 80, 22]:

$$R_{\text{fin}} \cong (I_{AB} - I_{AE}) \frac{I'_{AB}}{I_{AB}} R_{\text{raw}} . \qquad (6.3)$$

$I_{AB} = 1 + D \log_2 D + (1 - D) \log_2 (1 - D)$, $I_{AE} \cong 0.3 + I_{2\nu}$ and $I'_{AB} = 1 + D \log_2 D - \frac{7}{2} D$, where $D = QBER$. Here $I_{2\nu}$ is due to multi-photon pulses and has values of about 0.06, 0.14, and 0.40 for 5, 10, and 20 dB losses [76].

Security proofs are briefly alluded: For the BB84 protocol the security proof GLLP against an arbitrary attack exists [32]. Under weaker assumptions the NSG proof applies [61]. For the so-called SARG protocol a proof exists for incoherent attacks [9]. For the decoy-state protocol a security proof against an arbitrary attack has also been published [32, 49]. In Chap. 5 some attack strategies on QKD protocols are presented.

6.2.2 One-Way Weak Coherent Pulse QKD, Phase Coding (Toshiba)

In this section the one-way decoy pulse QKD system is discussed employing a protocol which involves one-way decoy pulses together with vacuum pulses [19].

QKD affects the secure communication between two remote parties Alice and Bob where the security of the keys is determined by the laws of quantum mechanics rather than the use of strong, one-way mathematical functions of encryption [27]. Since the original proposal [6] there has been an amount of work, beginning with the first experimental demonstration in 1992 [5], but reliable and compact systems compatible with existing telecom fiber technology are now starting to emerge [76] [29].

In the ideal case the QKD setup should be designed employing a true single photon source to guarantee immunity against the so-called photon number splitting (PNS) attacks from a potential eavesdropper (Eve) [83, 49, 12]. However, there is a lack of deterministic and reliable single photon sources. Most of the implementations use heavily attenuated lasers which emit photon pulses with a Poissonian number distribution. The PNS attack consists of blocking true single photons in the quantum channel and removing part of the multi-photon pulses by transmitting the remaining portion to Bob. Eve can then determine all or part of the key [12]. For further information on PNS attacks cf. Sect. 5.3.1.

In 2003 Hwang proposed to circumvent the PNS attack using additional (decoy) pulses sent by Alice [42]. The idea was to intersperse the signal pulses randomly with some "decoy pulses" that are weaker on average and so very rarely contain a multi-photon pulse. If Eve attempts a PNS attack, she will therefore transmit a lower fraction of the decoy pulses to Bob than the signal pulses. Thus, by monitoring the transmission of decoy and signal pulses separately, the attack can be detected [83] [49]. This means that stronger pulses may be used securely. A proof of the decoy pulse protocol has been given which also includes realistic experimental assumptions (GLLP) [32]. Recently a promising one-way QKD system was presented employing a single decoy pulse [85].

For now it is instructive to describe the method of phase coding in two interferometers (shown in Fig. 6.2). In the BB84 protocol Alice prepares randomly four

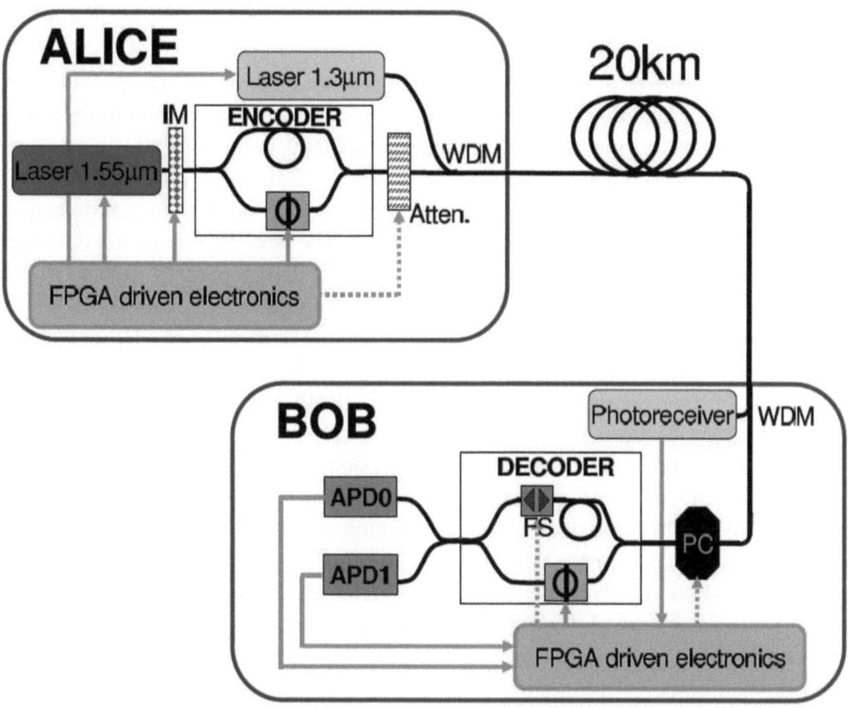

Fig. 6.2 Sketch of the optical layout of the one-way weak coherent pulse QKD system (phase-coding). The system represents a BB84 phase encoding protocol including weak + vacuum decoy states. Atten.: attenuator, IM: intensity modulator, PC: polarization controller, WDM: wavelength division multiplexer, FS: fiber stretcher, APD: avalanche photo diodes, FPGA: field programmable gate array

states using a first interferometer where the two arms have different lengths in order to produce sufficient time delays between the pulses. In one of the arms a phase shifter α is inserted. The phase shift Φ in Fig. 6.2 at Alice's box represents the quantity α used in the formalism here. The principle of the BB84 protocol can easily be understood for the idealized case where a single input photon (state $|1\rangle|0\rangle$, including the vacuum port $|0\rangle$), is given. Inside the interferometer four states are generated which belong to two mutually orthogonal bases $\frac{1}{\sqrt{2}}(\imath|1\rangle|0\rangle + e^{\imath\alpha}|0\rangle|1\rangle)$ where $\alpha = 0, \pi$ (bits 0 and 1 in the X basis) or $\alpha = \frac{\pi}{2}, \frac{3\pi}{2}$ (bits 0 and 1 in the Y basis). Time separation (or – equivalently – space separation of the pulses) is not included in the formalism in order to simplify matters. For time separation a complete wave packet description of the phase-coding protocol BB84 is necessary [78]. In short, the delay lines at Alice and Bob have to be equal in order to have interference between photons which take, e.g., in the first interferometer the short path and in the second interferometer the long path or vice versa. Only these events are indistinguishable from a quantum mechanical point of view [59].

But let's continue our prior considerations. The state behind Alice's interferometer can be expressed as $\frac{1}{2}[(e^{\imath\alpha} - 1)|1\rangle|0\rangle + \imath(e^{\imath\alpha} + 1)|0\rangle|1\rangle]$ describing the two outputs [25]. The probabilities of the two outputs are $\sin^2(\frac{\alpha}{2})$ and $\cos^2(\frac{\alpha}{2})$, respectively. If a coherent state $|\sqrt{\mu}\rangle|0\rangle$ is considered as input, the two outputs behind the interferometer can be described by the product state $|\frac{\sqrt{\mu}}{2}(e^{\imath\alpha} - 1)\rangle|\frac{\imath\sqrt{\mu}}{2}(e^{\imath\alpha} + 1)\rangle$ where μ is the mean photon number of the pulse to be considered. The probabilities of the outcomes are in this case $\mu\sin^2(\frac{\alpha}{2})$ and $\mu\cos^2(\frac{\alpha}{2})$. Bob, receiving the corresponding state, has a similar interferometer and detects in the X(Y) basis by phase shift $\beta = 0(\beta = \frac{\pi}{2})$. The phase shift Φ in Fig. 6.2 at Bob's box represents the quantity β used in the formalism here. His interferometer has two output detectors. If, e.g., he sets $\beta = 0$ and Alice has taken $\alpha = 0$ or $\alpha = \pi$, one of his detectors obtains a conclusive result which determines bit 0 or 1 (basis X). In case of $\beta = 0$ and $\alpha = \frac{\pi}{2}$ or $\frac{3\pi}{2}$ the detectors of Bob click by chance [Alice (Y) and Bob (X) use different bases]. A complementary process happens for $\beta = \frac{\pi}{2}$. Having consistent bases Alice and Bob retain their data while discarding the other ones. This completes the process of data sifting in the protocol BB84. Both bases correspond thus to an interferometric measurement.

In Fig. 6.2 a one-way fiber-optic QKD system with phase encoding is used. Two Mach–Zehnder phase encoding interferometers are applied. Alice and Bob are linked by a 20 km fiber spool through which the signal (an optical pulse with wavelength $\lambda = 1.55\,\mu\text{m}$) is transmitted at a repetition rate of about 7 MHz. The clock pulses ($\lambda = 1.3\,\mu\text{m}$), which do not overlap the signal pulses, have a duration of 5 ns each and deploy as synchronization. An intensity modulator is used in order to produce signal and decoy pulses of different intensities at random times whereas vacuum decoy pulses are produced by omitting trigger pulses to the signal laser. The signal and decoy pulses are strongly attenuated to the single photon level, while a strong clock pulse is then multiplexed with them to provide synchronization. Bob's detectors are two single photon InGaAs avalanche photodiodes (APD's).

The weak coherent pulse (WCP) decoy state + vacuum state BB84 protocol mentioned above was implemented [42]. The mean number of photons per pulse for signal and decoy states has to be chosen to be $\mu = 0.55$ and $\nu = 0.10$, respectively. The optimal probabilities of the various pulses are signal $N_\mu = 0.93$, decoy $N_\nu = 0.06$, vacuum $N_0 = 0.01$.

The properties of the detectors are carefully adjusted in order to avoid so-called fake-state attacks [55, 54, 56] and time-shift attacks [66, 86]. For further details on these kinds of attacks cf. Sect. 5.3.3 and 5.3.4, respectively.

A secure bit rate of greater than 10 kbps over 60 h was observed. This is approximately two orders of magnitude higher than what can be achieved at a fiber distance of 20 km without decoy states [30]. It is assumed that such a system could be very useful to be placed in a real-world environment such as a quantum network with fiber links of around a few tens of kilometers.

It should be mentioned that phase-coding QKD was demonstrated using an optically excited, triggered single photon source (SPS) emitting at a wavelength of $\lambda = 1.3\,\mu\text{m}$ [44]. The SPS (quantum dot source) shows a tenfold reduction in multi-photon emission compared to a laser and has been used to distribute keys secure from the PNS attack over 35 km along an optical fiber [32].

6.2.3 Coherent One-Way System, Time Coding (COW)

The coherent one-way QKD system COW was developed by GAP (Group of Applied Physics, University of Geneva) [75, 28, 74]. A sketch of the configuration is drawn in Fig. 6.3. The COW protocol described below is based to a certain extent on the well-known BB84 protocol [6, 27].

In the previous section the BB84 protocol has been described. Two mutually orthogonal bases, X and the Y, have been applied. However, a third basis $\{|1\rangle|0\rangle, |0\rangle|1\rangle\}$, called Z, can be used in principle, where applying this basis means

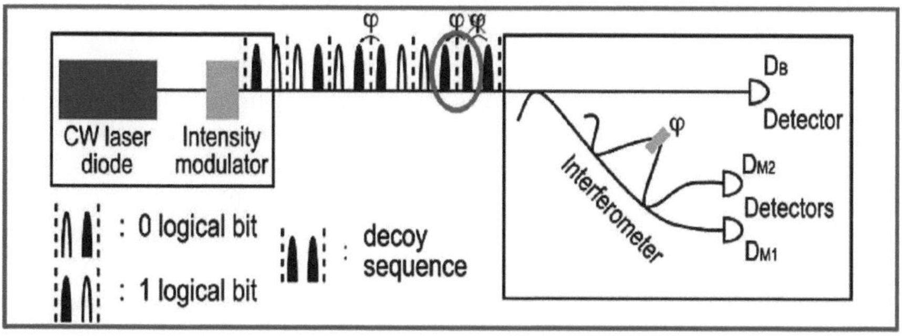

Fig. 6.3 Conceptual scheme of the quantum channel for implemented BB84 protocol with time-coding (coherent one-way-system COW); the left (*right*) box belongs to Alice (Bob); the transmission coefficient t_B to Bob's detector D_B amounts to 0.9, the transmission coefficient for the interferometer line is $(1 - t_B)$ and has a value of 0.1; see text for details

simply measuring the time-of-arrival of photons and is thus insensitive to optical errors [16]. This concept will be established in the following COW protocol where the Y basis is replaced by the Z basis, the X basis being used only occasionally to check coherence [48].

Now Fig. 6.3 comes into play. Alice's CW laser diode (1550 nm) together with the intensity modulator either prepares pulses of coherent states of mean photon number μ or completely blocks the beam (empty or "vacuum" pulses). The source pulses leaving Alice's laboratory with encoded information form a sequence of coherent states. In particular, the protocol exploits the coherence between any pair of subsequent pulses. The kth logical bits 0 and 1 are encoded in two-pulse sequences which can be written in each case as a product of a coherent state $|\sqrt{\mu}\rangle$ and an empty state $|0\rangle$ leading to

$$|0_k\rangle = |\sqrt{\mu}\rangle_{2k-1}|\sqrt{0}\rangle_{2k} \ , \quad |1_k\rangle = |\sqrt{0}\rangle_{2k-1}|\sqrt{\mu}\rangle_{2k} \ . \tag{6.4}$$

Note that in Eq. (6.4) the states $|0_k\rangle$ and $|1_k\rangle$ are not orthogonal because coherent states are not orthogonal by nature. A time-of-arrival measurement, whenever conclusive, provides the optimal unambiguous determination of the bit value [28]. To check coherence, a fraction $f \ll 1$ of *decoy sequences* is produced which can be written as a tensor product $|\sqrt{\mu}\rangle_{2k-1}|\sqrt{\mu}\rangle_{2k}$. Now due to the very narrow-banded CW laser including a large coherence length there is a well-defined phase between any two non-empty pulses within both, each *decoy sequence* but also across the *bit separation* [a $(1-0)-$ bit sequence]. Since equally spaced pulses are produced, the coherence of both, *decoy* and *(*$1-0$*)*- bit sequences, can be checked with a single interferometer. Eve cannot count the number of photons in any finite number of pulses without introducing errors: photon number splitting (PNS) attacks can be detected [28]. This is in contrast to the BB84 protocol where PNS attacks are repelled by a decoy-state technique, which consists in varying μ [42, 83, 49].

The pulses propagate to Bob through a quantum channel characterized by a transmission t and are split at a $[t_B : (1-t_B)]-$ beam splitter with a transmission coefficient $t_B \lesssim 1$ (e.g. $t_B = 0.9$). Only about 10% of the pulses are reflected into Bob's interferometer (*monitoring line*) which is destined to check quantum coherence. The pulses that are transmitted with about 90% (*data line*) are used to establish the raw key by measuring the times of arrival. The counting rate at detector D_B is $R = 1 - e^{-\mu t t_B \eta} \approx \mu t t_B \eta$, where η is the quantum efficiency of the photon counter (between 5 and 15%) and $\mu \approx 0.1-0.5$. Dark counts and limited efficiency of the detectors have to be taken into account.

The phase shift φ in Bob's interferometer is adjusted in such a way that within the time slots (dashed lines in Fig. 6.3) only detector D_{M1} clicks for regular two-pulse sequences (logical bits 0 or 1). When both pulses are non-empty [*decoy sequence* or a $(1-0)$-bit sequence] detector D_{M2} can fire as well at a time where only D_{M1} should have fired [78] resulting in a reduced visibility (both detectors fire). Here coherence can be quantified by Alice and Bob through the visibility

$$V = \frac{p(D_{M1}) - p(D_{M2})}{p(D_{M1}) + p(D_{M2})}, \tag{6.5}$$

where $p(D_{Mj})$ is the probability that detector D_{Mj} fires. These probabilities are small, the average detection rate on the monitoring line being $\propto \mu t(1 - t_B)\eta$ per pulse. However, if the bit rate is high, enough counting rates can be achieved in a reasonable time. The interferometer is only used to estimate the information of the eavesdropper and cannot introduce errors on the key.

The system is tolerant to reduced visibility and robust against PNS attacks. The system is insensitive to polarization fluctuations in fiber. One other advantage is that standard telecom components can be used.

The protocol can be summarized as follows:

1. Alice sends a large number of sequences of "logical bit 0" and "logical bit 1," both with probability $\frac{1-f}{2}$, and "*decoy sequences*" with probability $f \ll 1$.
2. Bob, after reception of the sequences, reveals for which bits he obtained detections on D_B in the *data line* (raw key) and when the detector D_{M2} has fired in the *monitoring line*.
3. Alice tells Bob which bits he has to remove from his raw key, since those bits are due to the detections of *decoy sequences* (sifting).
4. Alice analyzes the detections in Bob's detector D_{M2}. She estimates the break of coherence through the visibilities $V_{(1-0)}$ (coherence across a bit separation) and V_d (*decoy sequences*) and computes Eve's information. It is considered as a part of the protocol that Alice and Bob reject the key unless $V_d = V_{(1-0)}$.
5. Alice and Bob run error correction and privacy amplification ending up with a secret key.

Proving the security of the COW protocol remains a work in progress. The standard methods for proving the security of QKD protocols were, so far, developed for protocols in which the quantum symbols are sent one by one (e.g., qubits in the BB84, B92, six-state protocols [27]). The COW protocol, however, does not use this symbol-per-symbol type of coding and the standard security proofs do not apply in any straightforward way. To the contrary, the COW protocol is a so-called distributed-phase-reference protocol (such as also the "differential-phase-shift" protocol [43]) that relies on the overall phase, or more specifically, the coherence between successive non-empty pulses, to ensure the security of the protocol. For this family of protocols, much is still to be done at the level of security proofs.

So far, the security of the COW protocol has been proven against some particular attacks, such as the beam-splitting attack (BSA) [28, 75, 10, 11] and some intercept-resend (IR) attacks [28, 75]. Proofs for more specific attacks, such as zero-error attacks based on unambiguous state discrimination (USD) [10] and (in the limit of high losses) against 1-pulse or 2-pulse attacks that generalize the BSA by introducing errors [11], have also been shown (see Chap. 5).

6.2.4 Continuous Variables with Gaussian Modulation, QKD with Coherent States (CV)

The designed and realized prototype that implements a continuous variable (CV) QKD protocol is based on coherent states and reverse reconciliation [21, 2]. This is a stable and automatic system working constantly during 57 h and yielding an average secret key distribution rate of 8 kbits/s over 15 km standard optical fiber, including all quantum and classical communication.

So-called CV protocols rely on both quadratures of a coherent state and have been proposed in [67, 39, 14, 37, 34]. A complete implementation of the Gaussian-modulated coherent-state reverse reconciliation (RR) CV-QKD protocol is described in [50]. In the protocol, the quadratures operators of position $\hat{x} = \frac{1}{\sqrt{2}}(\hat{a}^+ + \hat{a})$ and momentum $\hat{p} = \frac{\imath}{\sqrt{2}}(\hat{a}^+ - \hat{a})$ of a train of coherent-state pulses are modulated in the complex plane with a Gaussian modulation of variance $V_A N_0$ at Alice's side, where N_0 is the shot noise variance that appears in the Heisenberg relation $\Delta x \Delta p \geq N_0$. The quantities $\hat{a}$ and $\hat{a}^+$ are the bosonic annihilation resp. creation operators defined for a coherent state $|\alpha\rangle$ using the Eigenvalue equation $\hat{a}|\alpha\rangle = \alpha|\alpha\rangle$. The Eigenvalue $\alpha = |\alpha|e^{\imath\theta} = (x + \imath p)$ is a complex number. These coherent states are sent from Alice to Bob through the quantum channel, along with a strong phase reference (phase angle θ_{LO}) – or strong local oscillator (LO) expressed classically by the complex number $\alpha_{LO} = |\alpha_{LO}|e^{\imath\theta_{LO}}$. Bob randomly measures the $\hat{x}$ or $\hat{p}$ quadrature by making the signal interfere with the LO in a pulsed, shot noise-limited homodyne detector. The phase shifting φ rotates the quadratures $\hat{x}$ and $\hat{p}$ to $\hat{x}_\varphi = \hat{x}\cos\varphi + \hat{p}\sin\varphi$ and $\hat{p}_\varphi = -\hat{x}\sin\varphi + \hat{p}\cos\varphi$, and it can be shown easily [25] that balanced homodyne detection (measuring the difference of the output intensities $\hat{I}_{21}$ behind a 50/50 beam splitter) means measuring the rotated quadrature component: $\hat{I}_{21} \propto |\alpha_{LO}|\hat{x}_\varphi$. A balanced homodyne detection is an amplifier. The LO amplifies the signal by the mutual optical mixing of the two. Or, seen from a different point of view, the homodyne detection is an interferometer that can be measurably imbalanced by a weak signal mode, because the reference field (LO) is very intense.

We now discuss the CV-QKD system [21]. As shown in Fig. 6.4. Alice produces coherent light pulses of 1550 nm wavelength with a frequency of $\omega = 500\,kHz$; the time between two pulses amounts therefore to $\Delta t = 1/\omega = 2\,\mu$s. The length of the generated pulses is 100 ns. The pulses are separated into a weak signal (path above) and a strong LO (path below) using an asymmetric 99/1 coupler. The LO contains typically 10^7 photons per pulse. Both the signal and the LO have to be polarized in the same direction. The signal pulses are then displaced in the complex plane, with arbitrary amplitude $|\alpha|$ and phase θ, randomly chosen from a two-dimensional Gaussian distribution centered at zero and with an adjustable variance $V_A N_0$. Alice's intended modulation variance is adjusted with a second amplitude modulator. Time and polarization multiplexing are used so that S and LO are transmitted to Bob in the same optical fiber without interfering. A 2×40 m delay line ($\doteq$ time delay $\Delta t' = \frac{80\,n}{c} = 400$ ns, where $n = \frac{3}{2}$ is the index of refraction, and c is the speed of

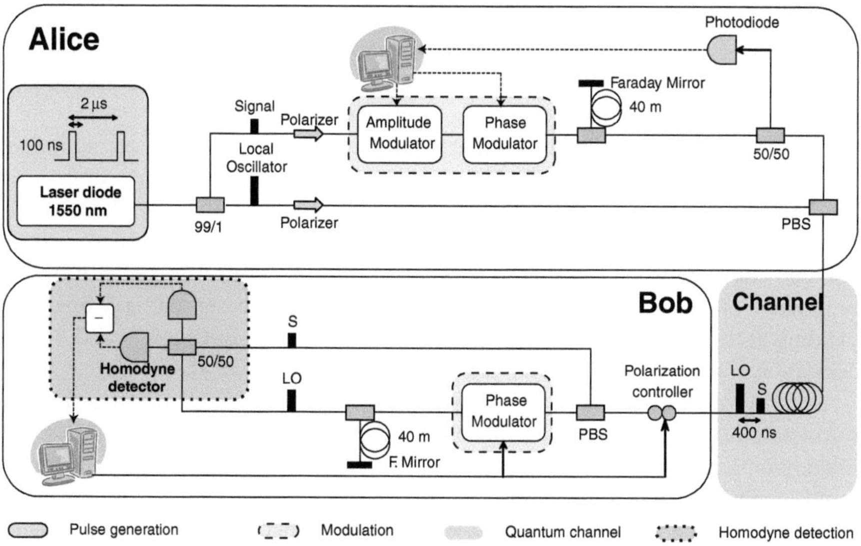

Fig. 6.4 Sketch of the optical layout of the CV-QKD prototype; LO: local oscillator, S: signal, PBS: Polarizing beam splitter; see text for details

light in vacuum) is inserted using a PBS (because the signal is polarized) and including a Faraday mirror (FM). The FM consists of a standard mirror and a 45° rotator. It therefore reflects the signal pulse by imposing a 90° rotation on its polarization. This system eliminates therefore all birefringence-induced polarization drifts. Using a 50/50 beam splitter, Alice can detect and store the Gaussian-modulated coherent-state signal S. Because of the orthogonal polarizations, the signal pulses (which contain about 10^2 photons after attenuation) and the LO pulses are then coupled in the transmission fiber using a PBS. The signal and LO pulses leave the PBS therefore at the same output port and travel through the quantum channel with orthogonal polarizations, being delayed by 400 ns.

In Bob's system, the LO pulse is transmitted through a PBS and a phase modulator imposes randomly a phase shift $\theta_{LO} = 0$ or $\theta_{LO} = \frac{\pi}{2}$. Using an analog delay line as at Alice's side the LO pulses are delayed by 400 ns and polarization rotated by 90°. Having now the same polarization, S and LO hit simultaneously the 50/50 beam splitter of a homodyne detection device where they interfere. The homodyne detection system outputs an electric signal, whose intensity is proportional to the rotated quadrature $\hat{x}_\varphi$ of the signal, where $\varphi = (\theta - \theta_{LO})$ is the phase difference between S and LO. Bob measures randomly either $\hat{x}_\theta$ and $\hat{x}_{\theta-\pi/2}$ to select one of the two quadratures. Later, using a public authenticated channel, he informs Alice about which quadrature he measured, so she may discard the irrelevant data. After many similar exchanges, Alice and Bob share a set of correlated Gaussian variables, which are called "key elements."

On Bob's side, homodyne detection is performed by choosing at random one of the two relevant quadrature measurements. Consider then a simple detection scheme, in which bit values are assigned by the sign of the detection signal, + or −, with respect to the half planes in the quantum optical phase space. As a result, both sender and receiver have binary data at hand. Classical data processing is then necessary for Alice and Bob to obtain a fully secret binary key.

Classical data processing means that Alice and Bob apply an error correction (or reconciliation) algorithm to their data. In order for Alice to correct the errors that appear in her data with respect to Bob's data (reverse reconciliation), the two parties perform a multilevel reconciliation process based on low-density parity check (LDPC) codes which are described in detail in [50]. To extract the secret information, Alice and Bob use privacy amplification algorithms based on hash functions yielding at the end a shorter bit sequence unknown to Eve. In Chaps. 3 and 4 error correction and privacy amplification are discussed in more detail.

Security proofs against individual Gaussian attacks including direct or reverse reconciliation are treated in the literature [37, 36, 34, 35]. Using Shannon's formula, simple analytical expressions for a Gaussian channel are derived. Security proofs against general collective attacks can be found in [60] and [24] with Shannon information being replaced by the Holevo quantity. Unconditional security against coherent attacks has also been proven [33, 69].

The secret key rate R is given as the difference between the amount of information shared by Alice and Bob and the amount of information available to Eve. For the system under consideration, the following results have been obtained [50, 21]:

$$I_{AB} = \frac{1}{2}\log_2 \frac{V + \chi_{\text{tot}}}{1 + \chi_{\text{tot}}}, \tag{6.6}$$

I_{AB}, mutual information between Alice and Bob
$V = (V_A + 1)$, excess noise ($V \approx 18.5$ shot noise units N_0)
V_A, Gaussian modulation of variance at Alice's side
$\chi_{\text{tot}} = (\chi_{\text{line}} + \chi_{\text{hom}}/T)$, total noise added between Alice and Bob
$\chi_{\text{line}} = (1/T - 1 + \epsilon)$, channel noise referred to the channel input
$T = 10^{-\alpha l/10}$, transmission coefficient of the channel
α, attenuation of the line ($= 0.2$ dB/km)
l, length of the line (e.g., 15 km $\rightarrow T = 0.5$)
ϵ, excess noise at the channel's input ($= 0.005$ shot noise units N_0)
$\chi_{\text{hom}} = (1 + v_{el})/\eta - 1$, detector's noise referred to Bob's input
v_{el}, noise added by the detection electronics ($= 0.041$ N_0 units)
η, detector efficiency ($= 0.606$).

Now two different attacks are taken into account yielding different key rates. These are (a) individual attacks which entail the Shannon rate and (b) collective attacks which give rise to the Holevo rate.

(a) Shannon rate (individual attacks): In the individual attack, Eve performs her measurement after Bob reveals his quadrature measurement (sifting) but before the error correction. Thus, her information is restricted to the Shannon information.

In the reverse reconciliation protocol Eve's information reads

$$I_{BE} = \frac{1}{2}\log_2\left[\frac{T\,(V+\chi_{\mathrm{tot}})}{\frac{1}{T\,(1/V+\chi_{\mathrm{line}})}+\chi_{\mathrm{hom}}}\right], \tag{6.7}$$

and the Shannon raw key rate reads

$$R^{Sh} = \beta\, I_{AB} - I_{BE}\,. \tag{6.8}$$

β is a factor which takes into account the efficiency of reconciliation algorithm ($\beta = 0.898$ [50]).

(b) Holevo rate (collective attacks): Eve's accessible information after error correction is now upper bounded by the Holevo quantity [68]. One obtains

$$R^{\mathrm{Hol}} = \beta\, I_{AB} - \chi_{BE}\,. \tag{6.9}$$

χ_{BE}, the Holevo information bound, reads as follows:

$$\chi_{BE} = G\left(\frac{\lambda_1-1}{2}\right) + G\left(\frac{\lambda_2-1}{2}\right) - G\left(\frac{\lambda_3-1}{2}\right) - G\left(\frac{\lambda_4-1}{2}\right), \tag{6.10}$$

$G(x) = (x+1)\log_2(x+1) - x\log_2 x$,
$\lambda_{1,2} = (1/2)[A \pm \sqrt{A^2-4B}]\,, \lambda_{3,4} = (1/2)[C \pm \sqrt{C^2-4D}]$,
$A = V^2(1-2T) + 2T + T^2(V+\chi_{\mathrm{line}})^2$, $B = T^2(V\chi_{\mathrm{line}}+1)^2$,
$C = \frac{V\sqrt{B}+T(V+\chi_{\mathrm{line}})+A\chi_{\mathrm{hom}}}{T(V+\chi_{\mathrm{tot}})}$, $D = \sqrt{B}\frac{V+\sqrt{B}\chi_{\mathrm{hom}}}{T(V+\chi_{\mathrm{tot}})}$.
Example: $l = 25$ km ($T = 0.30$) $\rightarrow R^{Sh} = 15.2$ kb/s and $R^{\mathrm{Hol}} = 12.3$ kb/s.

6.2.5 Entanglement-Based QKD (EB)

The use of quantum entanglement in combination with existing fiber telecommunication networks extends the possibility to implement long-distance QKD communication in the future [18].

The most important quotation related to the entanglement-based QKD system (polarization encoding) described here is given in [40]. A schematic picture of some technical details of the system is shown in Fig. 6.5.

The source emits polarization-entangled photons. The state can be written as

$$|\Phi\rangle = \frac{1}{\sqrt{2}}(\,|\,H_{810}\,H_{1550}\,\rangle + |\,V_{810}\,V_{1550}\,\rangle\,)\,. \tag{6.11}$$

Here H (V) indicates horizontal (vertical) polarization of a photon. The signal-mode photon (wavelength 1550 nm) and the idler-mode photon (wavelength 810 nm) are generated in pairs by an asymmetric SPDC (spontaneous parametric down-

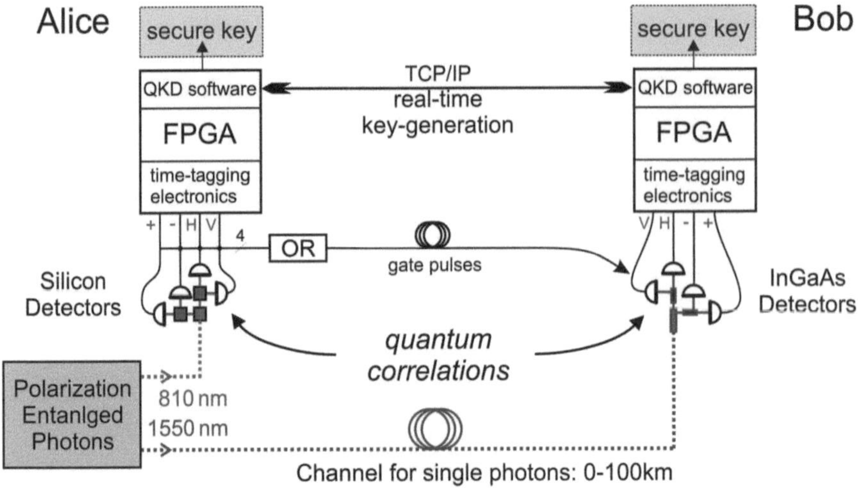

Fig. 6.5 Schematic picture of the entanglement-based QKD system. The source emits polarization-entangled photons of different wavelengths: 810 and 1550 nm. The 810 nm pair-photons are polarization analyzed locally by Alice and detected by her silicon detectors (APDs, avalanche photo diodes). The partner photons at 1550 nm are transmitted to Bob via telecom fibers on spool, analyzed and detected by InGaAs-APDs. Classical communication is carried out via a TCP/IP (transmission control protocol/internet protocol). FPGA: field programmable gate array; see text for details

conversion) process [70, 63, 46] activated by two quasi-phase matched periodically, poled $KTiOPO_4$ $(ppKTP)$ crystals which have been tailored from a 532 nm pump laser for type-I collinear generation of the photons (the pump laser and the crystals are not shown in the picture).

The reason for choosing the 1550/810 set of wavelengths is twofold: Because of the low absorption the signal is transmitted very well over optical fibers (1550 nm), but has to be registered using InGaAs detectors which must be gated. Therefore, the second pair photon (idler) is generated at a wavelength of 810 nm, which is detectable more easily with Si-APDs. This photon is immediately measured at Alice (at the source) with a passive BB84 detector module described in the next section in more detail. Then the outcomes are correlated with the measurement results at Bob's 1550 nm-detection module in both bases. The implemented protocol is called BBM92 protocol and is a version of BB84 for entanglement-based systems [7].

The description of the system shown in Fig. 6.5 goes as follows: The source at Alice described above produces asymmetric entangled photons which are separated by wavelength and coupled into single-mode fibers. The polarization analyzer for 810 nm photons is a free-space unit with a passive basis choice between measurements in the H/V or $+/-$ basis. This basis choice is realized by a first 50/50 beam splitter. The polarization measurements are performed with two polarizing beam splitters (PBSs). The 810 nm photons are detected in a four-channel Si-APD array. A delay module can set individual delay values for each Si-APD eliminating side channel attacks based on timing differences on Alice's side. Each detection event is recorded with a time stamp and an optical pulse (trigger) is generated at 1610 nm. The trigger signal is wavelength multiplexed onto the quantum channel operating at 1550 nm.

The quantum channel is made up of a single-mode standard telecom fiber.

At Bob's side the quantum signal is wavelength multiplexed to separate the quantum signal at 1550 nm from the trigger at 1610 nm. Electronic polarization controller is used to correct for polarization drifts in the quantum channel. Classical polarization state analysis is performed using polarimeters both on Alice's and on Bob's side (not shown in Fig. 6.5). A fiber beam splitter acts as basis choice for the BB84 polarization analysis at Bob. Two fiber-based PBSs split the incoming light into two orthogonal polarization components. Four InGaAs detectors measure the photon in each of the four BB84 polarization states. Trigger signals are converted into gate pulses on Bob's FPGA board. Delay lines guarantee that the four InGaAs detectors at Bob's side open their gate at the same time.

Coincidences of detector clicks between Alice and Bob establish a secret key after classical communication over TCP/IP.

Long-distance transmission of EB polarization-encoded qubits was successfully demonstrated in free space up to 144 km [82]. In long-distance fiber-based EB communication [57, 79] chromatic dispersion (CD) and birefringence (because of polarization-mode dispersion, PMD) [15] seem to play an important role. A similar EB QKD scheme described above was used to demonstrate the successful distribution of secret keys over 1.45 km fiber glass in an urban environment where the system was installed between the headquarters of a large bank and the Vienna city hall in order to accomplish a bank transfer [65].

Security proofs for EB systems, in which the source is under Eve's control, and where higher losses can be tolerated if the source is in the middle between Alice and Bob rather than if it is on Alice's side, are presented in [53]. If deviations from a perfect two-photon source are observed, EB security proofs take the presence of multi-photon components into account when error rates are considered [45]. Technological imperfections of single photon detectors produce nonobservable increase in the quantum bit error rate QBER (fake-state attacks [55, 54, 56] and time-shift attacks [66, 86]).

The goal of the QKD system is to estimate a final secret bit rate R. In the EB system discussed above the formula for the achievable secret key fraction reads [3] [81, 4](SECOQC project [58])

$$R = \frac{1}{2} \nu_A \, p_D \, [\, 1 \; - \; H(\delta) \; - \; f(\delta) \; H(\delta) \,] \; . \tag{6.12}$$

The particular quantities have the following meaning:
$\frac{1}{2}$, this factor stems from the BB84 sifting procedure
ν_A, Alice's detection rate, $\nu_A \approx \eta_A t_A \mu \approx 10^6 \, s^{-1}$
η_A, quantum efficiency of Alice's detector (50%)
t_A, transmittance in Alice's side (40%)

μ, optimal pair generation ($\mu \approx 5.10^6 s^{-1}$)
p_D, measured probability of Bob's detection ($p_D \approx 2p_{dB} + \eta_B t_B t$)
p_{dB}, probability of dark count per gate at Bob's detector ($= 3.10^{-5}$)
η_B, quantum efficiency of Bob's detector (10%)
t_B, transmittance in Bob's side (40%)
t, line transmittance with d being a distance ($t = 10^{-\alpha d/10}$)
α, fiber attenuation ($\alpha = 0.25\,\text{dB/km}$)
$H(\delta) = -\delta \log_2(\delta) - (1-\delta)\log_2(1-\delta)$, Shannon entropy
$f(\delta)$, depends on implemented error correction algorithm ($= 1.1$)
δ = *QBER*, measured quantum bit error rate ($0.5\% \lesssim \delta \lesssim 3.5\%$)

δ is a function of t, p_D, PMD, the two-photon visibility V and of multi-pair contributions. Neglecting dark counts on Alice's side the approximate formula for the error rate reads [4]

$$\delta \approx \frac{1}{p_D}\left[p_{dB} + \left(\frac{1-V}{2} + \frac{\mu \Delta T}{2} + \beta \sqrt{\frac{d}{d_0}} \right) \eta_B \, t_B \, t \right]. \tag{6.13}$$

V, two-photon visibility ($= 99\%$)
$\frac{\mu \Delta T}{2}$, multi-pair contribution to the error rate
ΔT, width of the gate pulse at Bob's detector ($= 2\,\text{ns}$)
$\beta\sqrt{\frac{d}{d_0}}$, formula for PMD [27]
$\beta = 0.001$ and $d_0 = 25\,\text{km}$, two parameters, experimentally found
d, distance Alice–Bob in kilometer

For $d = 0\,\text{km} \rightarrow \delta \approx 0.5\,\% \rightarrow R \approx 20\,\text{kHz}$. If we take the distance between Alice and Bob to be $d = 17\,\text{km}$ one obtains a quantum bit error rate $\delta \approx 1.3\%$ and a key rate $R \approx 6\,\text{kHz}$.

6.2.6 Free-Space QKD (FS)

Quantum cryptography is not necessarily restricted to the transmission of the quantum state by means of fiber glass. Free-space QKD has become an accepted method which has some advantages compared to the systems discussed so far. Interestingly enough, the first laboratory demonstration of QKD by Bennett and Brassard in 1989 was a free-space experiment over 30 cm air using the polarization state of photons for the generation of a quantum key. This system is described in detail in [5, 6]. In the meanwhile one is able to execute QKD over large free-space distances [73, 72]. Here we report on a free-space QKD implementation which uses strongly attenuated laser pulses over a distance of 500 m [84].

The system corresponds to the BB84 protocol [6] where qubits are encoded in the polarization of faint laser pulses. This protocol can be described as follows: Alice prepares single photons with randomly chosen polarization of four non-orthogonal states (horizontal (H), vertical (V), +45° or −45°). She sends the photons to Bob, who analyzes the polarization in a randomly and independently chosen basis (either H/V or ±45°). Subsequently, Alice and Bob compare publicly their basis choices while discarding events where they had chosen different bases (sifting).

Quantum mechanics does not allow an eavesdropper (Eve) to measure the polarization of a single photon without introducing errors between the communicating parties. The quantum bit error rate (QBER) of the sifted key gives an upper bound on the information Eve might have gained. This quantity is calculated during the classical error correction procedure and at the same time is needed to make sure that the information of a potential eavesdropper on the key is negligible (privacy amplification).

There have been some free-space experiments over relatively large distances [73, 41, 47] showing the possibility to build global quantum key exchange systems based on quantum communication satellites [64]. Below, a QKD free-space system is described which can be applied to shorter distances in urban areas [84]. In Fig. 6.6 a sketch of the elements of such a system is drawn. Two units can be identified: the left side belongs to the transmitter unit (Alice) and the right side belongs to the receiver unit (Bob). The units are discussed subsequently.

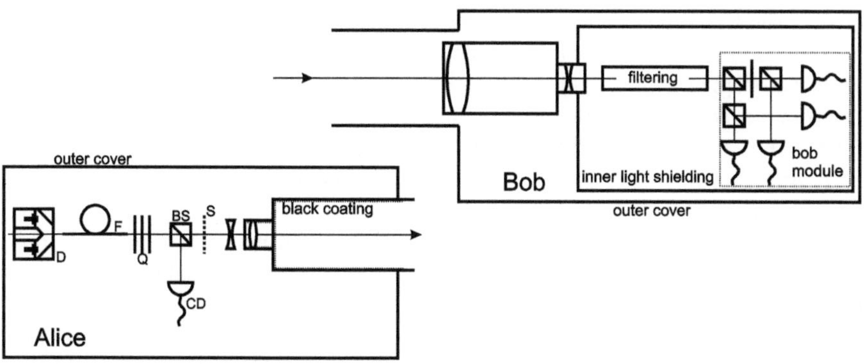

Fig. 6.6 Simplified diagram of the free-space Alice and Bob setup: D: cube, housing the laser-diodes, F: fiber modefilter, Q: quarter- and half-wave plates for polarization compensation of the fiber, BS: beam splitter, CD: detector for calibration of mean photon numbers, S: shutter to prevent daylight coming in when calibrating the mean photon numbers. The front lens of the Alice telescope is protected from sunlight by a 25 cm long black coated tube. Therefore the receiver collects less stray light. Bobs unit see text

The transmitter unit consists of four laser diodes together with a conical mirror and a spatial filter. The diodes produce in random sequence (according to the choices of basis and bit values) weak coherent pulses of polarized light (polarization: H, V, +45°, −45°) of mean photon number μ of 0.1 approximately. The advantage of this method is that no active polarization manipulation is needed. The four laser diodes emitting 850 nm wavelength are seen on the left side in Fig. 6.6. They are arranged around a conical mirror where the four beams are reflected and combined into one direction. The beams have to pass a spatial filter in order to let pass only a single spatial mode and are then injected into a single-mode fiber. A monitoring APD is mounted behind a BS. The second output of the BS leads to a telescope which emits the photons into free space.

To ensure that as many photons from Alice as possible are detected by Bob, two telescopes are employed, one at each end. Detailed data can be achieved elsewhere [73]. In [84], a distance of about 500 m between two buildings in downtown Munich has been chosen.

The receiver unit (Bobs module, right side of Fig. 6.6) is directly attached to the end of the receiver telescope. A non-polarizing BS, a set of two PBSs, and a HWP are used to perform the polarization analysis of the incoming photons. The 50/50 BS selects the basis (H/V or $\pm 45^{\circ}$) randomly and the subsequent PBS including the HWP determines whether a conclusive polarization measurement can be obtained. The module acts completely passive. The single photons are detected by four silicon APDs. Attached to the outputs of Bob's module is the time stamp unit that records the time-of-arrival of each detection event (timing resolution better than 1 ns).

Once the synchronization task is finished, Alice and Bob can start the key sifting process. If Alice has sent the photon in the same basis as measured by Bob, they will use the assigned bit value for the sifted key. Ideally, both sifted keys would be perfectly correlated. But experimental imperfection and/or an eavesdropper cause errors. The CASCADE algorithm is used for error correction [38]. Public communication is necessary during this procedure. Each publicly transmitted bit increases the eavesdropper's knowledge of the key. Hence, these announced bits have to be taken into account during the privacy amplification process [52]. In Chaps. 3 and 4 CASCADE and privacy amplification are treated more precisely.

A sifted key of about 50 kbit/s (QBER=3.2%) and a secure key of about 17 kbit/s could be achieved during night operation. However, weather conditions affect the attenuation of the free-space link. It has been observed that only fog, heavy rain, or snowfall limits the transmission severely as well as strong turbulence above sun-heated roofs close to the optical path.

Recently, a paper has been published presenting an entanglement-based QKD system overcoming the limitation of the night operation. Here spectral, spatial, and temporal filtering techniques are used in order to achieve a secure key continuously over several days under varying light and weather conditions [62].

6.2.7 Low-Cost QKD (LC)

A free-space quantum-cryptographic system that can operate in daylight conditions is described in the following [17]. The system acts at a wavelength of 632.8 nm. It works over a short range of a few meters and is designed to eventually work in short-range consumer applications. The free-space QKD system for short distances should be a low-cost application that is aimed at protecting consumer transactions. The design philosophy is based on a future hand-held electronic credit card using free-space optics. A method is proposed to protect these transactions using the shared secret stored in a personal hand-held transmitter. Thereby Alice's module is integrated within a small device such as a mobile telephone, or PDA (personal digital assistant), and Bob's module consists of a fixed device such as a bank ATM (a-synchrone transfer mode).

Quantum cryptography provides a means for two parties to securely generate shared messages. The information is encoded in non-orthogonal quantum states which an eavesdropper cannot measure without disturbing. The most commonly used method is devised on the BB84 protocol [6]. It is based on polarization encoding presented in the preceding section. This method is also applied in the low-cost QKD system described here. After the sifting procedure Alice and Bob should theoretically be correlated unless eavesdropping has taken place on the quantum channel. The eavesdropper must measure in a random polarization basis uncorrelated to that of Alice and will therefore inject errors. This should be actively monitored by measuring the error rate and discarding data where the error rate exceeds a certain threshold (typically $\sim 12\%$).

In an experimental setup errors like optical imperfections, background counts, and detector noise limit the significance of the results. Therefore, the processes of error correction [13] and privacy amplification [8] are performed resulting in a secret key shared only by Alice and Bob.

Long-distance free-space QKD experiments have been developed over 23 km [47] and 10 km [41] just as well as studies in order to develop QKD systems between ground and low earth orbit satellites [31].

Alice's module uses a driver circuit which produces sub-5 ns pulses. The driver pulses are combined with the output from a digital input/output card (NuDAQ) and passed to four AlInGaP LEDs, see Fig. 6.7. The NuDAQ card is regulated by an external clock and passes a random bit string, generated by a quantum number generator (QRNG, idQuantique), to the Alice module, recording which LED fires. The driver pulses are produced at a repetition rate of 5 MHz. The four LEDs are attached to a holder with dichroic sheet polarizers which are orientated in each of the four polarization states, 0°, 90°, 45° and −45°, placed over each output. Not till then the light is polarized. To combine the four beam paths, a diffraction grating is used. Because of the special design the holder serves to direct the polarized light toward

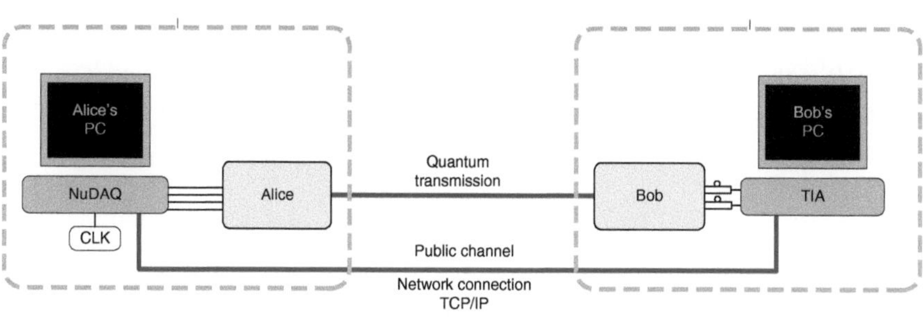

Fig. 6.7 Schematic diagram of a short-distance free-space QKD system. NuDAQ: digital input/output data acquisition card, CLK: external clock, TIA: time interval analyzer card, TCP/IP: transmission control protocol/internet protocol; see text for details

the grating. A pinhole is placed after the grating together with a 50 mm focal lens to collimate the beam. A 632.8 ± 3 nm filter is included to limit the bandwidth.

Currently the free-space quantum transmission between Alice and Bob consists of a fixed connection of roughly 5 cm distance. In addition, Alice and Bob communicate classically via the internet which is a real public channel.

Bob's module consists of a setup similar to Alice's module but in reversed order. First the beam coming from Alice is filtered and focused to a similar grating, where the beam is diffracted to four detectors with dichroic sheet polarizers (orientated in the four polarization directions) over each detector. Four passively quenched silicon avalanche photodiodes are used for detection. A simple circuit takes the output from the detectors and converts it to a readable positive pulse. Time-of-arrival information is recorded by a time interval analyzer card (TIA).

The bit error rate is $BER = N_{\text{wrong}}/N_{\text{total}}$ where N_{wrong} is the number of erroneous bits and N_{total} is the number of bits received in total. The module has to be able to operate in daylight conditions, so the background error rate [31] is the most limiting factor of the entire system.

The signal count S for this system is

$$S = \frac{R\,\nu\,T\,\eta}{4} \tag{6.14}$$

including the following quantities:

R, pulse repetition rate (5 MHz)
ν, expected photon number (~ 0.3 [52])
T, lumped transmission including geometric losses (~ 1, because of the short-range system)
η, efficiency of the detection system (~ 0.045)

The protocol effectively splits the signal into four on the detectors, leading to a factor 4 reduction in signal bit rate after sifting the key. The background rate is given by

$$P_b = B\,t\,, \tag{6.15}$$

where B is the background count rate per detector ($B \leq 36{,}000$ counts/s) and t is the time synchronization gate (5 ns). The background error rate for this system is [17]

$$E = 0.027 + \frac{P_b}{S}\,. \tag{6.16}$$

The current version of Bob's module can operate in shaded areas but not in full direct sunlight.

Most QKD systems use the CASCADE algorithm [38] for error correction, which we discussed in Sect. 3.2. Another method has been adapted here. The so-called low-density Parity Check (LDPC) algorithm [23] is implemented for the system in the case under consideration. The advantage of this code is that the protocol has very little interactive communication compared to CASCADE. The error corrected keys are then passed through the privacy amplification process. This effectively reduces the length of the key. To estimate the key length, the Lütkenhaus bound is used [51]. Currently the system is able to establish more than 4000 secret bits per second between transmitter and receiver in low-light conditions.

In the next generation device it is expected to be able to operate at 10,000 secret bits per second up to full daylight conditions. Eventually, Alice's module must be able to be brought to Bob's module and fully automatically aligned. This can be done using a fixed alignment cradle or docking station.

6.3 Summary

Quantum key distribution has triggered intense and prolific research work during the past 10 years and now progresses to maturity. During the EC/IST project SECOQC the seven most important QKD systems have been developed or refined, and this work compiled to some extent the physical and technical principles developed so far. These systems are (1) the plug and play system, (2) the Phase-Coding QKD, (3) the time-coding QKD, (4) the continuous variables system, (5) the entanglement-based system, (6) the free-space system, and (7) the low-cost QKD.

(1) The Plug and Play auto-compensating system designed by idQuantique SA employs a strong laser pulse ($\lambda = 1550\,\text{nm}$) emitted by Bob's laser diode. The pulse is separated at a first 50/50 BS. The two pulses travel down to the two input ports of a PBS, after having traveled, respectively, through a short and a long arm which includes a phase modulator and a delay line. The linear polarization is turned by 90° in the short arm, so that the two pulses exit the PBS by the same port. The separated pulses travel down to Alice, are reflected on a FM, attenuated, and come back orthogonally polarized. In turn, both pulses now take the other path at Bob's interferometer and arrive simultaneously at the first BS where they interfere. They are detected by InGaAs APD's. Since the two pulses follow the same path in the interferometer (short–long or long–short), the system is auto-compensated. The BB84 and the SARG protocols are implemented using phase coding. Alice applies one of the four phase shifts on the second pulse of each pair and Bob's phase modulator completes the protocol. In order to avoid noise enhancement by elastic Rayleigh scattering, the laser pulses are emitted in trains and are stored in a long delay line at Alice's side before being sent back to Bob. The Plug and Play auto-compensating design offers the advantage of being highly stable and passively aligned.

(2) The one-way weak pulse system (phase-coding) designed by Toshiba Research Europe Ltd. is a fiber-optic, decoy-state system with phase encoding. It employs a decoy protocol using weak and vacuum pulses. It uses two asymmetric Mach-Zehnder interferometers for encoding and decoding. Both the signal and the decoy

pulses are generated by a 1550 nm pulsed laser diode. The pulses are modulated with an intensity modulator to generate the necessary ratio of signal pulse strength to decoy pulse intensity and are then strongly attenuated to the desired level before leaving Alice's side. An active stabilization technique is used for continuous operation.

(3) The coherent one-way system (time-coding) designed by GAP-Université de Genève, idQuantique SA and ARC realizes the novel distributed-phase-reference COW protocol. Here Alice's CW laser diode (1550 nm) and an intensity modulator prepares pulses of weak coherent states or completely blocks the beam. A time-of-arrival measurement and an interferometer at Bob's side provide both optimal unambiguous determination of bit values and check of coherence for signal and decoy sequences using a phase shifter. The visibility of the various pulse sequences behind the interferometer provides information about an attack of an eavesdropper.

(4) The continuous variables system developed by CNRS-Institute d'Optique-Univ. Paris-Sud, THALES Research and Université Libre de Bruxelles implements a coherent-state reverse-reconciliated QKD protocol. The protocol encodes the key information on both quadratures of the electromagnetic field. Alice uses a pulsed laser diode and an asymmetric BS to generate signal and local oscillator pulses. The signal pulses are appropriately modulated in amplitude and phase. The signal is then time- and polarization-multiplexed with the delayed local oscillator before propagating through the quantum channel. After de-multiplexing, Bob uses a time-resolved homodyne detection system in order to measure the quadrature selected by his phase modulator. Alice and Bob share therefore correlated continuous data. Efficient low-density parity check error correction and privacy amplification provide for generating a binary secret key.

(5) The entanglement-based QKD system was developed by an Austrian–Swedish consortium (University of Vienna, ARC and Royal Institute of Technology of Kista). The system uses the unique quantum mechanical property of entanglement for transferring the correlated measurements into a secret key. The asymmetric source at Alice produces two photons of different wavelengths (810 nm and 1,550 nm) by spontaneous parametric down-conversion in a ppKTP crystal set. The 810 nm photon is measured by Alice in a passive system (one 50/50 BS and two PBSs, four Si detectors), thus implementing the BBM92 protocol for entangled states. The other 1,550 nm photon travels down the quantum channel with low-transmission losses and is registered by Bob using a similar passive system except for the four InGaAs detectors. The quantum correlations between the photons create a secret key.

(6) The free-space QKD system was developed by the group of H. Weinfurter from the university of Munich. It employs the BB84 protocol using polarization-encoded attenuated laser pulses with photons of 850 nm wavelength. Decoy states are used to ensure key security even with faint pulses. The system is applicable to day and night operation using excessive filtering in order to suppress background light. The transmitter (Alice) applies four laser diodes producing random sequences of weak coherent pulses of different polarization. The photons travel down a free-space distance between two telescopes (at Alice's and Bob's side) and are registered

by the receiver (Bob). The receiver mainly consists of the spatial filter and the polarization analyzing module with single-photon detectors.

(7) The low-cost QKD system was developed by John Rarity's team of the university of Bristol. The system can be applied for secure banking including consumer protection. The system can operate in daylight conditions and uses the BB84 protocol. The transmitter (Alice) consists of a hand-held computer interfaced to a FPGA. Pulses from the FPGA drive four polarized LEDs which are combined in a diffractive optical element. The receiver (Bob) is run from a laptop computer. His optical system uses a standard BS-based analyzer with four photon detectors. The main error source is background light. Error correction is accomplished by the low-density parity check code followed by privacy amplification to ensure security.

References

1. Alléaume (Editing author), R., Bouda, J., Branciard, C., Debuisschert, T., Dianati, M., Gisin, N., Godfrey, M., Grangier, P., Länger, T., Leverrier, A., Lütkenhaus, N., Painchault, P., Peev, M., Poppe, A., Pornin, T., Rarity, J., Renner, R., Ribordy, G., Riguidel, M., Salvail, L., Shields, A., Weinfurter, H., Zeilinger, A.: Quantum key distribution and cryptography. SECOQC White Paper (2007)
2. Assche, G., Cardinal, J., Cerf, N.: IEEE Trans. Inf. Theory **50**, 394 (2004)
3. Beaudry, N., Moroder, T., Lütkenhaus, N.: arXiv:0804.3082 (2008)
4. Beaudry, N., Moroder, T., Lütkenhaus, N.: in preparation (2008)
5. Bennett, C., Bessette, F., Brassard, G., Salvail, L., Smolin, J.: J. Cryptol. **5**, 3 (1992)
6. Bennett, C., Brassard, G.: Conf. on Computers, Systems and Signal Processing, Bangalore, India (IEEE, New York) p. 175 (1984)
7. Bennett, C., Brassard, G., Mermin, N.: Phys. Rev. Lett. **68**, 557 (1992)
8. Bennett, C., Brassard, G., Robert, J.M.: SIAM J. Comp. **17(2)**, 210 (1988)
9. Branciard, C., Gisin, N., Kraus, B., Scarani, V.: Phys. Rev. A **72**, 032,301 (2005)
10. Branciard, C., Gisin, N., Lütkenhaus, N., Scarani, V.: Quant. Inf. Comput. **7**, 639 (2007)
11. Branciard, C., Gisin, N., Scarani, V.: New. J. Phys. **10**, 013,031 (2008)
12. Brassard, G., Lütkenhaus, N., Mor, T., Sanders, B.: Phys. Rev. Lett. **85**, 1330 (2000)
13. Buttler, W., Lamoreaux, S., Torgerson, J., Nickel, G., Donahue, C., Peterson, C.: Phys. Rev. A **67(5)**, 052,303 (2003)
14. Cerf, N., Lévy, M., Assche, G.: Phys. Rev. A **63**, 052,311 (2001)
15. Damask, J.: Polarization Optics in Telecommunications. **Springer**, New York (2005)
16. Debuisschert, T., Boucher, W.: Phys. Rev. A **70**, 1042,306 (2004)
17. Duligall, J., Godfrey, M., Harrison, K., Munro, W., Rarity, J.: New J. Phys. **8**, 249 (2006)
18. Dusek, M., Lütkenhaus, N., Hendrych, M.: Progress in Optics **49**, 381 (2006)
19. Dynes, J., Yuan, Z., Sharpe, A., Shields, A.: Opt. Express **15/13**, 8465 (2007)
20. Félix, S., Gisin, N., Stefanov, A., Zbinden, H.: J. Mod. Opt. **48**, 2009 (2001)
21. Fossier, S., Diamanti, E., Debuisschert, T., Villing, A., Tualle-Brouri, R., Grangier, P.: arXiv:0812.3292 (2008)
22. Fuchs, C., Gisin, N., Griffiths, R., Niu, C., Peres, A.: Phys. Rev. A **56**, 1163 (1997)
23. Gallager, R.: IRE Trans. Inf. Th. **IT-8**, 21 (1962)
24. García-Patrón, R., Cerf, N.: Phys. Rev. Lett. **97**, 190,503 (2006)
25. Gerry, C., Knight, P.: Introductory quantum optics. 1st edn. **Cambridge University Press**, Cambridge (2005)

26. Ghernaouti-Hélie (Editing author), S., Tashi, I., Länger, T., Monyk, C.: Quantum cryptography: An innovation in the domain of secure information transmission. SECOQC Business White Paper (2008)
27. Gisin, N., Ribordy, G., Tittel, W., Zbinden, H.: Rev. Mod. Phys. **74**, 145 (2002)
28. Gisin, N., Ribordy, G., Zbinden, H., Stucki, D., Brunner, N., Scarani, V.: arXiv:0411022 (2004)
29. Gobby, C., Yuan, Z., Shields, A.: Appl. Phys. Lett. **84**, 3762 (2004)
30. Gobby, C., Yuan, Z., Shields, A.: Electron. Lett. **40**, 1603 (2004)
31. Gorman, P., Rarity, J., Tapster, P., Knight, P.: New J. Phys. **4**, 82 (2002)
32. Gottesman, D., Lo, H.K., Lütkenhaus, N., Preskill, J.: Quant. Inf. Comp. **5**, 325 (2004)
33. Gottesmann, D., Preskill, J.: Phys. Rev. A **63**, 022,309 (2001)
34. Grosshans, F., Assche, G., Wenger, J., Brouri, R., Cerf, N., Grangier, P.: Nature **421**, 238 (2003)
35. Grosshans, F., Cerf, N.: Phys. Rev. Lett. **92**, 047,905 (2004)
36. Grosshans, F., Cerf, N., Wenger, J., Tualle-Brouri, R., Grangier, P.: Quant. Inf. Comp. **3**, 535 (2003)
37. Grosshans, F., Grangier, P.: Phys. Rev. Lett. **88**, 057,902 (2002)
38. Helleseth, T: (ed.). Advances in cryptology - proceedings of eurocrypt '93. Lecture Notes in Computer Science, Springer, Berlin **765**, 410 (1994)
39. Hillery, M.: Phys. Rev. A **61**, 022,309 (2000)
40. Hübel, H., Vanner, M., Lederer, T., Blauensteiner, B., Lorünser, T., Poppe, A., Zeilinger, A.: Opt. Express **15/12**, 7853 (2007)
41. Hughes, R., Nordholt, J., Derkacs, D., Peterson, C.: New J. Phys. **4(43)**, 1 (2002)
42. Hwang, W.Y.: Phys. Rev. Lett. **91**, 057,901 (2003)
43. Inoue, K., Walks, E., Yamamoto, Y.: Phys. Rev. Lett. **89**, 037,902 (2002)
44. Intallura, P., Ward, M., Karimov, O., Yuan, Z., See, P., Shields, A.: arXiv:0710.0565 (2007)
45. Koashi, M., Preskill, J.: Phys. Rev. Lett. **90**, 057,902 (2003)
46. Konig, F., Mason, E., Wong, F., Albota, M.: Phys. Rev. A **71**, 033,805 (2005)
47. Kurtsiefer, C., Zarda, P., Halder, M., Weinfurter, H., Gorman, P., Tapster, P., Rarity, J.: Nature **419**, 450 (2002)
48. Lo, H.K., Chau, H., Ardehali, M.: J. Cryptology **18**, 133 (2005)
49. Lo, H.K., Ma, X., Chen, K.: Phys. Rev. Lett. **94**, 230,504 (2005)
50. Lodewyck, J., Bloch, M., García-Patrón, R., Fossier, S., Karpov, E., Diamanti, E., Debuisschert, T., Cerf, N., Tualle-Brouri, R., McLaughlin, S., Grangier, P.: Phys. Rev. A **76**, 042,305 (2007)
51. Lütkenhaus, N.: Phys. Rev. A **59**, 3301 (1999)
52. Lütkenhaus, N.: Phys. Rev. A **61**, 052,304 (2000)
53. Ma, X., Fung, C.H., Lo, H.K.: Phys. Rev. A **76**, 012,307 (2007)
54. Makarov, V., Anisimov, A., Skaar, J.: Phys. Rev. A **74**, 022,313 (2005)
55. Makarov, V., Hjelme, D.: J. Mod. Opt. **52**, 691 (2005)
56. Makarov, V., Skaar, J.: Quant. Inf. Comput. **8**, 622 (2008)
57. Marcikic, I., de Riedmatten, H., Tittel, W., Zbinden, H., Legré, M., Gisin, N.: Phys. Rev. Lett. **93**, 180,502 (2004)
58. Monyk (Coordinator), C.: Development of a global network for secure communication based on quantum cryptography. EC/IST Integrated Project SECOQC **Contract No. 506813** (2004-2008)
59. Muller, A., Herzog, T., Huttner, B., Tittel, W., Zbinden, H., Gisin, N.: Appl. Phys. Lett. **70**, 793 (1997)
60. Navasqués, M., Grosshans, F., Acín, A.: Phys. Rev. Lett. **97**, 190,502 (2006)
61. Niederberger, A., Scarani, V., Gisin, N.: Phys. Rev. A **71**, 042,316 (2005)
62. Peloso, M., Gerhardt, I., Ho, C., Lamas-Linares, A., Kurtsiefer, C.: arXiv:0812.1880 (2008)
63. Pelton, M., Marsden, P., Ljunggren, D., Tenger, M., Karlsson, A., Fragemann, A., Canalias, C., Laurell, F.: Opt. Express **12**, 3573 (2004)

64. Pfennigbauer, M., Aspelmeyer, M., Leeb, W., Baister, G., Dreischer, T., Jennewein, T., Neckamm, G., Perdigues, J., Weinfurter, H., Zeilinger, A.: J. Opt. Netw. **4**, 549 (2005)
65. Poppe, A., Fedrizzi, A., Ursin, R., Böhm, H., Lorünser, T., Maurhardt, O., Peev, M., Suda, M., Kurtsiefer, C., Weinfurter, H., Jennewein, T., Zeilinger, A.: Opt. Express **12/16**, 3865 (2004)
66. Qi, B., Fung, C.H., Lo, H.K., Ma, X.: Quant. Inf. Comput. **7**, 73 (2007)
67. Ralph, T.: Phys. Rev. A **61**, 010,303(R) (1999)
68. Renner, R.: Ph.D. thesis **ETH Zürich** (2005)
69. Renner, R., Cirac, J.: arXiv:0809.2234 (2008)
70. Ribordy, G., Brendel, J., Gautier, J., Gisin, N., Zbinden, H.: Phys. Rev. A **63**, 012,309 (2000)
71. Scarani, V., Bechmann-Pasquinucci, H., Cerf, N., Dusek, M., Lütkenhaus, N., Peev, M.: arXiv:0802.4155 (2008)
72. Schmitt-Manderbach, T.: Diss., LMU München (2007)
73. Schmitt-Manderbach, T., Weier, H., Fürst, M., Ursin, R., Tiefenbacher, F., Scheidl, T., Perdigues, J., Sodnik, Z., Kurtsiefer, C., Rarity, J., Zeilinger, A., Weinfurter, H.: Phys. Rev. Lett. **98**, 010,504 (2007)
74. Stucki, D., Barreiro, C., Fasel, S., Gautier, J.D., Gay, O., Gisin, N., Thew, R., Thoma, Y., Trinkler, P., Vannel, F., H., Z.: arXiv:0809.5264 (2008)
75. Stucki, D., Brunner, N., Gisin, N., Scarani, V., Zbinden, H.: Appl. Phys. Lett. **87**, 194,108 (2005)
76. Stucki, D., Gisin, N., Guinnard, O., Ribordy, G., Zbinden, H.: New. J. Phys. **4**, 41.1 (2002)
77. Stucki, D., Ribordy, G., Stefanov, A., Zbinden, H., Rarity, J.: J. Mod. Opt. **48**, 1967 (2001)
78. Suda, M., Herbst, T., Poppe, A.: Eur. Phys. J. D **42**, 139 (2007)
79. Takesue, H.: Opt. Express **14**, 3453 (2006)
80. Tancevski, L., Slutsky, B., Rao, R., Fainman, S.: Proc. SPIE **3228**, 322 (1997)
81. Tsurumaru, T., Tamaki, K.: arXiv:0803.4226 (2008)
82. Ursin, R., Tiefenbacher, F., Schmitt-Manderbach, T., Weier, H., Scheidl, T., Lindenthal, M., Blauensteiner, B., Jennewein, T., Perdigues, J., Trojek, P., Ömer, B., Fürst, M., Meyenburg, M., Rarity, J., Sodnik, Z., Barbieri, C., Weinfurter, H., Zeilinger, A.: nature physics **doi**, 10.1038/nphys629 (2007)
83. Wang, X.B.: Phys. Rev. Lett. **94**, 230,503 (2005)
84. Weier, H., Schmitt-Manderbach, T., Regner, N., Kurtsiefer, C., Weinfurter, H.: Fortschr. Phys. **54/8-10**, 840 (2006)
85. Yuan, Z., Sharpe, A., Shields, A.: Appl. Phys. Lett. **90**, 011,118 (2007)
86. Zhao, Y., Fung, C.H., Qi, B., Chen, C., Lo, H.K.: arXiv:0704.3253 (2007)

Chapter 7
Statistical Analysis of QKD Networks in Real-Life Environment

K. Lessiak and J. Pilz

As discussed before, Quantum Key Distribution (QKD) has already been realized in various experiments. Due to this there is an interest to prove whether or not external influences like temperature, humidity, sunshine duration, and global radiation have an effect on the quality of QKD systems. In consequence there is also an interest to predict the qubit error rate (QBER) and the key rate (KR). In the course of the SECOQC project [10], measurements of different devices in the prototype network in Vienna have been conducted. Within these measurements the correlation between QBER, KR, and these environmental influences is analyzed. Therefore, statistical methods such as generalized linear models and generalized linear mixed models are used. Applying these two models predictions become possible. For the implementations we used the statistical software R© (R version 2.7.1).

In Sect. 7.1 the statistical methods used for the analysis are introduced. There the basic principles of generalized linear models (GLMs) are established. An introduction to generalized linear mixed models (GLMM) is also given. Furthermore, the results of the field experiment in Vienna are shown and processed in Sect. 7.2. Hereafter, in Sect. 7.3 the analyses and processing of the data take place.

7.1 Statistical Methods

Linear models play a decisive role in the economic, social, and engineering sciences, especially in the investigation of causal relationships. They establish a firm basis of statistical methodologies. Because of the different requirements in these sciences, a variety of models have been studied and made available. Very important extensions

K. Lessiak (✉)
Safety & Security Department, Quantum Technologies, AIT Austrian Institute of Technology GmbH, Lakeside B01A, 9020 Klagenfurt, Austria, katharina.lessiak@ait.ac.at; http://www.ait.ac.at

J. Pilz
Institute of Statistics, Klagenfurt University, Universitätsstraße 65-67, A-9020 Klagenfurt, Austria, juergen.pilz@uni-klu.ac.at

Lessiak, K., Pilz, J.: *Statistical Analysis of QKD Networks in Real-life Environment*. Lect. Notes Phys. **797**, 123–149 (2010)
DOI 10.1007/978-3-642-04831-9_7
© Springer-Verlag Berlin Heidelberg 2010

of linear models are, for example, generalized linear models, mixed effect models, and nonparametric regression models.

The starting point for these extensions is the linear regression model:

$$y = \beta_0 + \beta_1 x_1 + \cdots + \beta_p x_p + \varepsilon, \tag{7.1}$$

with response variable y and predictors x_1, ..., x_p. In this regression model ε is the error, which is assumed to be normally distributed.

In principle, there are three extensions of the linear model. The first extension of the above-mentioned regression model generalizes the x part. Basically, the predictors x in a linear model are combined in a linear way to model the effect on the response. But sometimes it is inadequate to capture the structure of the data in a linear way, because more flexibility is necessary. Additive models, trees, and neural networks are methods that allow a more flexible regression modeling of the response that combine the predictors in a nonparametric manner.

There also exist data which have a grouped, nested, or hierarchical structure and that lead to the second extension of the above-mentioned regression model, namely the generalization of the ε part. Because repeated measures, longitudinal, and manifold data consist of several observations taken on the same individual or group and the fact that this induces a correlation structure in the error ε, mixed effect models are needed.

In the following, the third extension is explained in detail, which generalizes the y part of the linear model.

7.1.1 Generalized Linear Models (GLMs)

GLMs have been introduced by Nelder and Wedderburn (1972) [7] and a detailed representation is given by McCullagh and Nelder (1989) [6]. GLMs are necessary to represent categorical, binary, and other response types because in standard linear models it is not possible to handle non-normal responses such as counts or proportions. Logit-, probit-, and Poisson models are important special cases of GLMs.

To define a GLM it is necessary to specify two components. At first, it is essential that the distribution of the response variable is a member of the exponential family. Furthermore, the link function shows how the mean of the response and a linear combination of the predictors are related.

Definition 7.1 (cf. [4]) In a GLM the distribution of Y is from the exponential family of distributions which take the general form

$$f(y \mid \theta, \phi) = \exp\left[\frac{y\theta - b(\theta)}{a(\phi)} + c(y, \phi)\right]. \tag{7.2}$$

The θ is called the canonical parameter and represents the location while ϕ is called the dispersion parameter and represents the scale.

The exponential family distributions have mean and variance (cf. [4]):

$$E(Y) = \mu = b'(\theta), \tag{7.3}$$

$$var(Y) = b''(\theta)\ a(\phi). \tag{7.4}$$

The mean is a function of θ only, while the variance is a product of functions of the location and the scale. $b''(\theta)$ is called the variance function and describes how the variance relates to the mean.

Let η denote the linear predictor:

$$\eta = \beta_0 + \beta_1 x_1 + \cdots + \beta_p x_p, \tag{7.5}$$

then the link function g describes how the mean response $E(Y) = \mu$ is linked to the covariates through the linear predictor:

$$\eta = g(\mu). \tag{7.6}$$

Thus, a specific generalized linear model is characterized by choosing the distributional type of the exponential family, the choice of the link function, and the definition of the predictors.

An example from the exponential family is the normal or Gaussian which is in the following defined by specifying the functions a, b, and c of Eq. 7.2 (cf. [4]):

$$\begin{aligned} f(y \mid \theta, \phi) &= \frac{1}{\sqrt{2\pi}\sigma} \exp\left[-\frac{(y-\mu)^2}{2\sigma^2}\right] \\ &= \exp\left[\frac{y\mu - \mu^2/2}{\sigma^2} - \frac{1}{2}\left(\frac{y^2}{\sigma^2} + \log(2\pi\sigma^2)\right)\right], \end{aligned} \tag{7.7}$$

with

$$\theta = \mu,\ \phi = \sigma^2,\ a(\phi) = \phi,\ b(\theta) = \theta^2/2, \text{ and } c(y, \phi) = -(y^2/\phi + \log(2\pi\phi))/2.$$

The parameter ϕ is free in contrary to other exponential family members as, for example, for the Poisson and binomial distributions. The reason is that the normal distribution is a two-parameter family. It is also important to state that in the normal case $b''(\theta) = 1$ and hence the variance is independent of the mean. The identity link in the normal distribution is defined as $\eta = \mu$.

Usually, the parameters β of a GLM can be estimated using maximum likelihood [6]. Furthermore, having estimated a GLM for a data set the goodness of fit has to be tested. In GLMs a measure for the goodness of fit is the deviance which measures how closely the observed values are approximated by the model-based fitted values of the response. To compare the deviance of two models a likelihood ratio test can be used. There a statistic is used, which has a χ^2-distribution with degrees of freedom equal to the difference in the number of parameters estimated under each model.

7.1.2 Generalized Linear Mixed Models (GLMMs)

The generalized linear mixed model (GLMM) is an extension of the GLM, especially to the context of clustered measurements. In a GLMM the idea of the GLM is combined with the idea to include random effects. So, in a GLMM the linear predictor $\eta = x^T\beta$ of the GLM is extended by adding random effects, which are responsible to account for the correlation structure of clustered observations. According to Faraway (2006) [4] the model takes the following form.

Definition 7.2 In a GLMM the response is a random variable Y taking values $y_1, \ldots, y_n$ and with a distribution coming from the exponential family as before,

$$f(y_i \mid \theta_i, \phi) = \exp\left[\frac{y_i\theta_i - b(\theta_i)}{a(\phi)} + c(y, \phi)\right]. \tag{7.8}$$

Moreover, with the canonical link $\theta_i = \mu_i$, the fixed effects β and the random effects γ are related to θ_i through

$$\theta_i = x_i^T\beta + z_i^T\gamma, \tag{7.9}$$

where x_i and z_i are the corresponding rows from the design matrices X and Z for the fixed and random effects. The random effects are assumed to follow a probability density $h(\gamma|V)$ with given (hyper-) parameters V.

A more specific definition for longitudinal and cluster data is given in Fahrmeir, Kneib, and Lang (2007) [3].

Furthermore, according to Faraway (2006) [4] for estimation the likelihood takes the form

$$L(\beta, \phi, V/y) = \prod_{i=1}^{n} \int f(y_i|\beta, \phi, \gamma)h(\gamma|V)d\gamma. \tag{7.10}$$

Due to the integral in the likelihood it becomes in general very difficult to compute this term explicitly. So, there are a lot of approaches to approximate the likelihood using theoretical or numerical methods. Following Verbeke and Molenberghs [13, 14] there are three main approximations to explain

- approximating the integrand
- approximating the data
- approximating the integral

The Laplace method is used for the approximation of the function to be integrated, a specific approach is given in Tierney and Kadane (1986) [12]. A penalized quasi-likelihood approach (PQL) can be used for "approximating the data" and is detailed in Wolfinger and O'Connel (1993) [15]. For the case that these methods fail (adaptive), Gaussian quadrature methods can be used, details are given in Pinheiro and Bates (1995) [8], (2000) [9]. Demidenko [2] gives an introduction to approximations of maximum likelihood by statistical simulation methods, especially Monte Carlo methods. Moreover, a Bayesian approach is possible, which is considered in Sinha (2004) [11] and a further interesting approach is the generalized estimating equations (GEE) considered by Zeger, Liang and Albert (1988) [17].

7.2 Results of the Experiments

As already described in Sect. 9.1, in the course of the SECOQC project in Vienna a prototype of a QKD network was implemented. There were seven QKD-link devices (described in Chap. 6) which connected five subsidiaries of SIEMENS Austria.

The measurements started on October 1, 2008 and were concluded on November 8, 2008. During this time the devices were tested and upgrades were executed. The data used in this chapter are based only on the measurements from the different devices from October 8, 2008 to October 10, 2008 of the quantum network because a lot of data gaps arise out of upgrades on the devices. During the measurements in the fiber ring network in Vienna data about the quality of QKD systems, especially the qubit error rate (QBER) and the key rate (KR), are obtained. The data on air temperature, humidity, sunshine duration, and global radiation are obtained from the Central Institute for Meteorology and Geodynamics (ZAMG), which is the national weather service agency in Austria. The calculations have been implemented in R and are based on [4] and [16].

7.2.1 Data Set for the Device "Entangled"

First of all the entangled-based device is discussed which is called "Entangled." Alice was positioned in the SIEMENS Austria location in Erdberg (ERD) and Bob was located in Siemensstrasse (SIE). The route ERD–SIE was used. The data set consists of qubit error rate (QBER), key rate (KR), temperature, humidity, sunshine duration, and global radiation. In Table 7.1 a short extract of the data set is presented.

Table 7.1 Extract of the data set of the device "Entangled"

QBER	KR	Temperature	Humidity	Sunshine duration	Global radiation
329	2734	148	88	0	0
345	2650	145	89	0	0
329	2728	143	90	0	0
336	3151	141	91	0	0
379	2489	140	91	0	0
354	3056	141	91	0	0

An initial graphical and numerical look at the data is essential for any data analysis. To get a compact numerical overview of the data the six statistics minimum, 1st quantile, median, mean, 3rd quantile, and maximum for each numerical

Table 7.2 Numerical overview of the results from the device "Entangled"

QBER		KR		Temperature	
Min. :	291.0	Min. :	1707	Min. :	117.0
1st Qu.:	345.0	1st Qu.:	2304	1st Qu.:	134.0
Median :	364.0	Median :	2559	Median :	148.0
Mean:	374.3	Mean:	2519	Mean:	149.5
3rd Qu.:	393.0	3rd Qu.:	2700	3rd Qu.:	163.0
Max. :	513.0	Max. :	3595	Max. :	184.0
Humidity		**Sunshine duration**		**Global radiation**	
Min. :	71.00	Min. :	0.0	Min. :	0.0
1st Qu.:	80.00	1st Qu.:	0.0	1st Qu.:	0.0
Median:	83.00	Median :	0.0	Median :	0.0
Mean :	84.07	Mean :	101.2	Mean :	92.98
3rd Qu.:	91.00	3rd Qu.:	0.0	3rd Qu.:	140.00
Max. :	93.00	Max. :	600.0	Max. :	539.00

variable are requested which are sufficient to get a rough idea of the distributions. In Table 7.2 these six statistics of the data set from the device "Entangled" are shown.

In Fig. 7.1 the scatterplot of the data set is shown. A high correlation between the variables is indicated by an elongated ellipse in the plot and it can be distinguished between a positive correlation and a negative correlation which depends on the orientation. It can be seen that QBER and KR are highly correlated. To get a numerical overview about the pairwise correlations the correlation matrix is useful. The correlation matrix of the device "Entangled" can be seen in Table 7.3.

The scatterplot matrix and the correlation matrix show that external influences like temperature, humidity, sunshine duration, and global radiation do not have an influence on the quality of QKD systems, i.e., on the entangled-based device. In contrast to the results of the measurements of April 23, 2008 shown in [5] an influence of these external influences on the quality of QKD systems can be observed. This difference is due to the fact that the devices were upgraded.

The next step to get a good overview and a good understanding of the data is to have a look on graphical summaries. An essential means to get an idea of the distribution of each numerical variable are histograms. A further well-known means is a density estimate. Boxplots are very useful to find out the outliers in the data set. In Fig. 7.2 the histogram, the density function, and the boxplot of QBER and KR are given. The boxplot of QBER shows a lot of outliers. Before testing the distributions it is important to eliminate these outliers. The reason is that these outliers can falsify the result.

To get further information about the distributions different tests are possible, for example, the Kolmogorov–Smirnov test. A good alternative is the Cramer–von Mises two sample test [1] because it is more powerful than the Kolmogorov–Smirnov test. The Cramer–von Mises two sample test declares that QBER can be assumed to be gamma distributed because the p-value is 0.36. The fact that the p-value is greater than 0.05 is an indication that this hypothesis cannot be rejected.

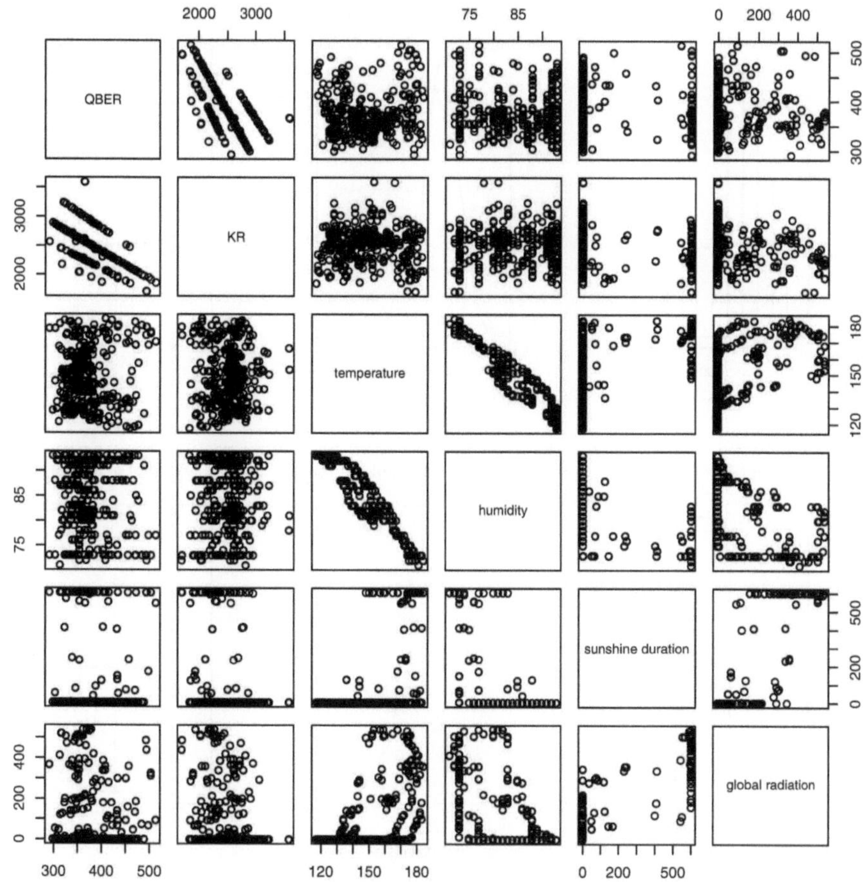

Fig. 7.1 Scatterplot matrix for the data set of the entangled-based device

Table 7.3 Correlation matrix for the device "Entangled"

	QBER	KR	Temperature
QBER	1.0000000	–0.7000019	0.0276082
KR	–0.7000019	1.0000000	–0.0179497
Temperature	0.0276082	–0.0179497	1.0000000
Humidity	–0.0824010	0.0895992	–0.9473179
Sunshine duration	0.0968474	–0.2794770	0.5800525
Global radiation	0.0075992	–0.2888295	0.5531215

	Humidity	Sunshine duration	Global radiation
QBER	–0.0824010	0.0968474	0.0075992
KR	0.0895992	–0.2794770	–0.2888295
Temperature	–0.9473179	0.5800525	0.5531215
Humidity	1.0000000	–0.6192554	–0.5692125
Sunshine duration	–0.6192554	1.0000000	0.8631108
Global radiation	–0.5692125	0.8631108	1.0000000

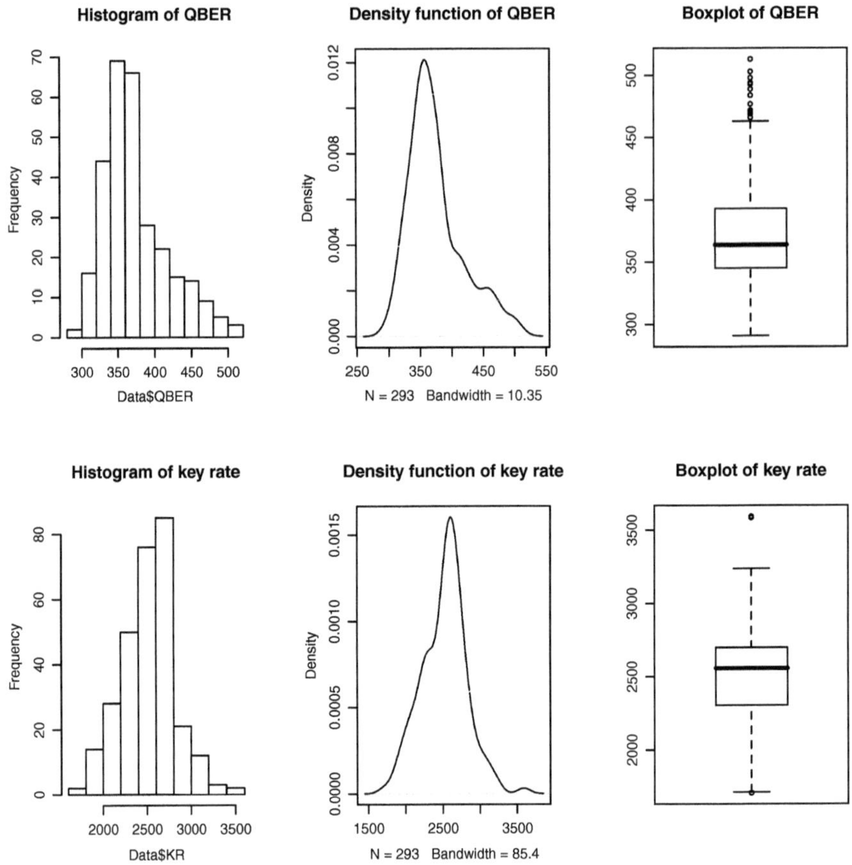

Fig. 7.2 Histogram, density function, and boxplot of QBER and KR of the device "Entangled"

7.2.2 Data Set for the Device "Freespace"

In the following, the data for the device "Freespace" are processed. As in Sect. 7.2.1 the measurements in the fiber ring network in Vienna started on October 8, 2008. The device "Alice" was positioned in Forum (FOR), the device "Bob" was located in ERD, and the route FOR–ERD was used. The data also consist of qubit error rate (QBER), key rate (KR), temperature, humidity, sunshine duration, and global radiation. In Table 7.4 a short extract of the data set is given.

Table 7.5 gives the six summary statistics. Furthermore, Fig. 7.3 shows the scatterplot for the data set "Freespace". In this scatterplot a correlation between QBER and KR can be seen. Moreover, the correlation matrix is given in Table 7.6.

The correlation matrix confirmed the correlation between QBER and KR. As in Sect. 7.2.1 the scatterplot matrix and the correlation matrix show that there is no linear dependence between air temperature, humidity, sunshine duration, and global

Table 7.4 Extract of the data set for the device "Freespace"

QBER	KR	Temperature	Humidity	Sunshine duration	Global radiation
203	17503	145	89	0	0
209	14065	143	90	0	0
215	11499	140	91	0	0
221	1906	141	91	0	0
220	13119	139	91	0	0
248	7697	132	92	0	0

Table 7.5 Numerical overview of the device "Freespace"

QBER		KR		Temperature	
Min. :	172.0	Min. :	952	Min. :	117.0
1st Qu.:	187.0	1st Qu.:	11969	1st Qu.:	133.2
Median :	208.0	Median :	14522	Median :	148.0
Mean:	218.2	Mean:	14455	Mean:	149.5
3rd Qu.:	222.0	3rd Qu.:	17428	3rd Qu.:	165.0
Max. :	849.0	Max. :	29302	Max. :	184.0
Humidity		**Sunshine duration**		**Global radiation**	
Min. :	71.00	Min. :	0.0	Min. :	0.0
1st Qu.:	78.25	1st Qu.:	0.0	1st Qu.:	0.0
Median:	84.00	Median :	0.0	Median :	0.0
Mean :	83.94	Mean :	104.2	Mean :	100.2
3rd Qu.:	91.00	3rd Qu.:	0.0	3rd Qu.:	153.0
Max. :	93.00	Max. :	600.0	Max. :	539.00

radiation on the one hand and QBER as well as KR on the other hand. Thus, it seems that external influences have no effect on the quality of QKD systems of the device "Freespace."

For a graphical overview, Fig. 7.4 displays the histogram, the density function, and the boxplot of QBER and KR. In the boxplot of QBER a lot of outliers can be seen. As aforementioned, it is important to eliminate these outliers because they can falsify the result, e.g., the prediction. It is particularly essential to eliminate them before testing the distributions of the numerical variables of the data set.

Alternatively, one could try to model the outlier-prone distribution of QBER on the basis of some extreme-value-distribution. This will not be considered here, however.

Again tests like the Kolmogorov–Smirnov test or the Cramer–von Mises two sample test are used to get further information about the distribution of QBER. The Cramer–von Mises two sample test supports the hypotheses that QBER of the data set is gamma, normal, and log-normal distributed, respectively, because in all cases the p-value is greater than 0.05. For the generalized linear mixed model (GLMM) in Sect. 7.3.2, QBER is taken as gamma distributed because the gamma distribution is more general than the normal or log-normal distribution.

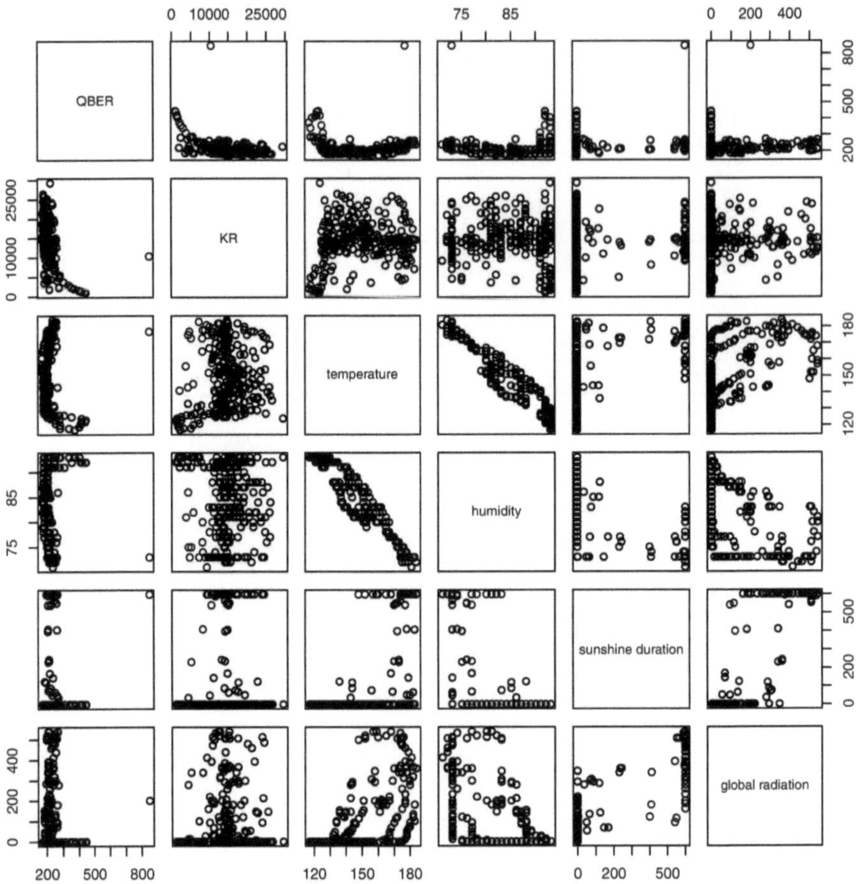

Fig. 7.3 Scatterplot matrix for the data set of the device "Freespace"

Table 7.6 Correlation matrix of the device "Freespace"

	QBER	KR	Temperature
QBER	1.0000000	–0.4637677	–0.0259311
KR	–0.4637677	1.0000000	0.1122507
Temperature	–0.0259311	0.1122507	1.0000000
Humidity	–0.0119615	–0.0997538	–0.9519479
Sunshine duration	0.1983148	–0.0006192	0.5981148
Global radiation	0.1222088	0.0179512	0.5844248

	Humidity	Sunshine duration	Global radiation
QBER	–0.0119615	0.1983148	0.1222088
KR	–0.0997538	–0.0006192	0.0179512
Temperature	–0.9519479	0.5981148	0.5844248
Humidity	1.0000000	–0.6266734	–0.5776959
Sunshine duration	–0.6266734	1.0000000	0.8419395
Global radiation	–0.5776959	0.8419395	1.0000000

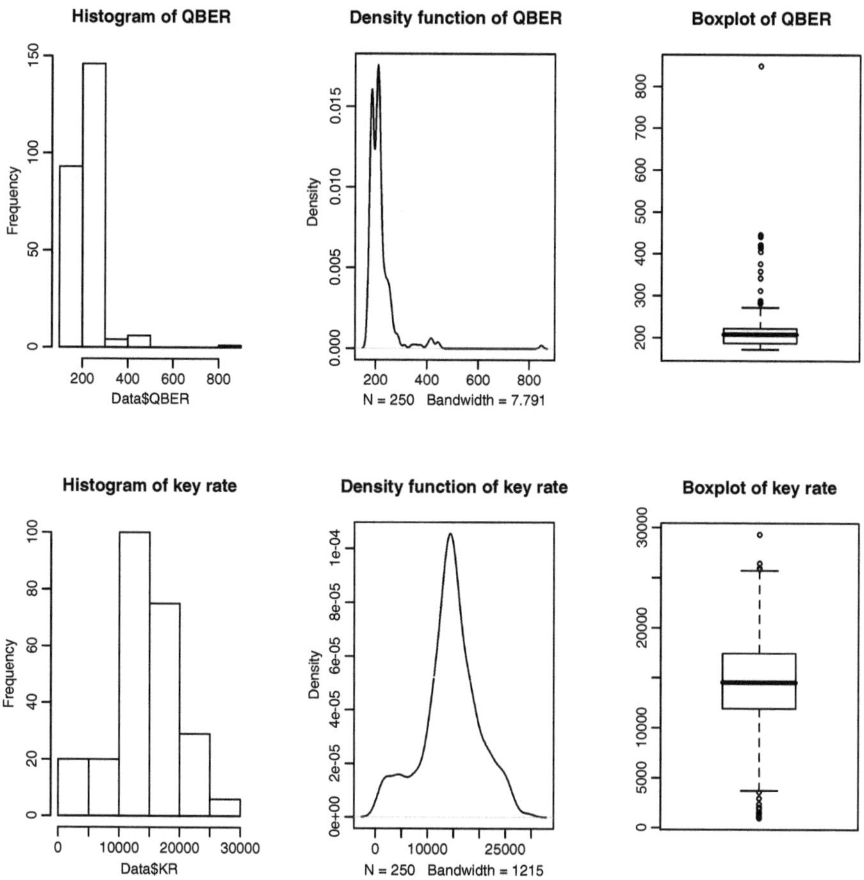

Fig. 7.4 Histogram, density function, and boxplot of QBER and KR of the device "Freespace"

7.2.3 Data Set for the Device "Autocompensating Plug&Play"

The measurements for the device "Autocompensating Plug&Play" in the fiber ring network in Vienna also started on October 8, 2008 where Alice was located in Gudrunstrasse (GUD) and Bob was positioned in Breitenfurterstrasse (BRT). The route GUD–BRT was used. In Table 7.7 a short extract of the data set is shown.

Table 7.7 Extract of the data set of the device "Autocompensating Plug&Play"

QBER	KR	Temperature	Humidity	Sunshine duration	Global radiation
146	1240	148	88	0	0
126	1250	145	89	0	0
150	986	143	90	0	0
146	1280	141	91	0	0
146	977	140	91	0	0
159	944	141	91	0	0

Table 7.8 gives the six summary statistics for QBER, KR, temperature, humidity, sunshine duration, and global radiation. In Fig. 7.5 the scatterplot matrix of the data set is given, similar to the data sets before a correlation between QBER and KR. For numerical information the correlation matrix is used, which is given in Table 7.9.

Table 7.8 Numerical overview of the device "Autocompensating Plug&Play"

QBER		KR		Temperature	
Min. :	98.0	Min. :	81.0	Min. :	117.0
1st Qu.:	133.0	1st Qu.:	434.0	1st Qu.:	135.0
Median :	148.0	Median :	512.0	Median :	148.0
Mean:	150.7	Mean:	525.2	Mean:	149.8
3rd Qu.:	165.0	3rd Qu.:	603.0	3rd Qu.:	163.5
Max. :	411.0	Max. :	1280.0	Max. :	184.0
Humidity		**Sunshine duration**		**Global radiation**	
Min. :	71.00	Min. :	0.0	Min. :	0.0
1st Qu.:	80.00	1st Qu.:	0.0	1st Qu.:	0.0
Median:	84.00	Median :	0.0	Median :	0.0
Mean :	83.99	Mean :	97.71	Mean :	94.02
3rd Qu.:	91.00	3rd Qu.:	0.0	3rd Qu.:	146.00
Max. :	93.00	Max. :	600.0	Max. :	539.00

The correlation matrix confirms the assumption that there is a correlation between QBER and KR. It can be seen that there is no correlation between QBER and temperature, humidity, sunshine duration, and global radiation. So, it can be assumed that external influences have no effect on the quality of QKD systems of the device type "Autocompensating Plug&Play."

In Fig. 7.6 the histogram, the density function, and the boxplot of QBER and KR are given. The next step is to find out the outliers in the data set and to eliminate them because they can lead to a faulty result of the statistical analysis. In the boxplot of the QBER three outliers can be clearly identified.

The Kolmogorov–Smirnov test declares that the QBER of the data set for the device "Autocompensating Plug&Play" is gamma, log-normal, and normally distributed, respectively, because the p-value in all three cases is greater than 0.05. The Cramer–von Mises two sample test confirmed the result of the Kolmogorov–Smirnov test because in all three cases the p-value is greater than 0.05, again, and for the same reason as before, the gamma distribution is assumed for QBER of the device "Autocompensating Plug&Play" – Sect. 7.3.2.

7.2.4 Data Set for the Device "Continuous Variables"

The measurements of the device "Continuous Variables" in the fiber ring network in Vienna also started on October 8, 2008. Alice was positioned in Erdberg (ERD) and Bob was located in Gudrunstrasse (GUD). For the measurements of the device "Continuous Variables" the route ERD – GUD was used. A short extract of the data set is given in Table 7.10.

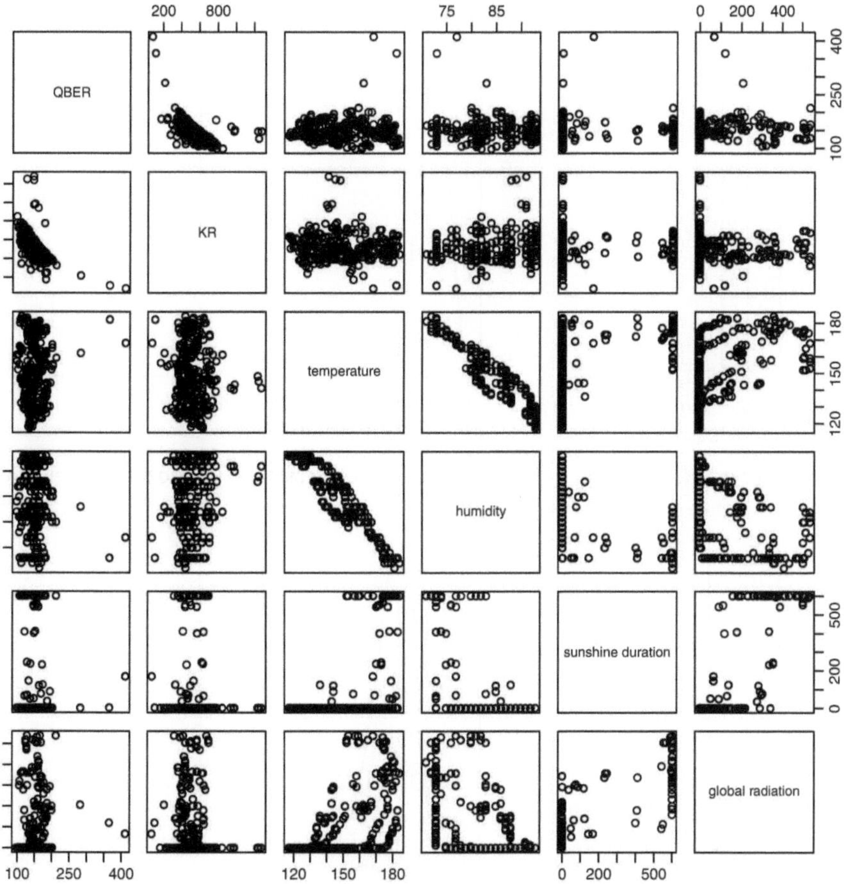

Fig. 7.5 Scatterplot matrix for the data set of the device "Autocompensating Plug&Play"

Table 7.9 Correlation matrix for the device "Autocompensating Plug&Play"

	QBER	KR	Temperature
QBER	1.0000000	–0.5974331	0.0664187
KR	–0.5974331	1.0000000	–0.0854596
Temperature	0.0664187	–0.0854596	1.0000000
Humidity	–0.0663922	0.1678327	–0.9487629
Sunshine duration	–0.0084388	–0.0543023	0.5785442
Global radiation	0.1105632	–0.1339746	0.5516755

	Humidity	Sunshine duration	Global radiation
QBER	0.0663922	–0.0084388	0.1105632
KR	0.1678327	–0.0543023	–0.1339747
Temperature	–0.9487629	0.5785442	0.5516755
Humidity	1.0000000	–0.6176290	–0.5626842
Sunshine duration	–0.6176290	1.0000000	0.8457442
Global radiation	–0.5626841	0.8457442	1.0000000

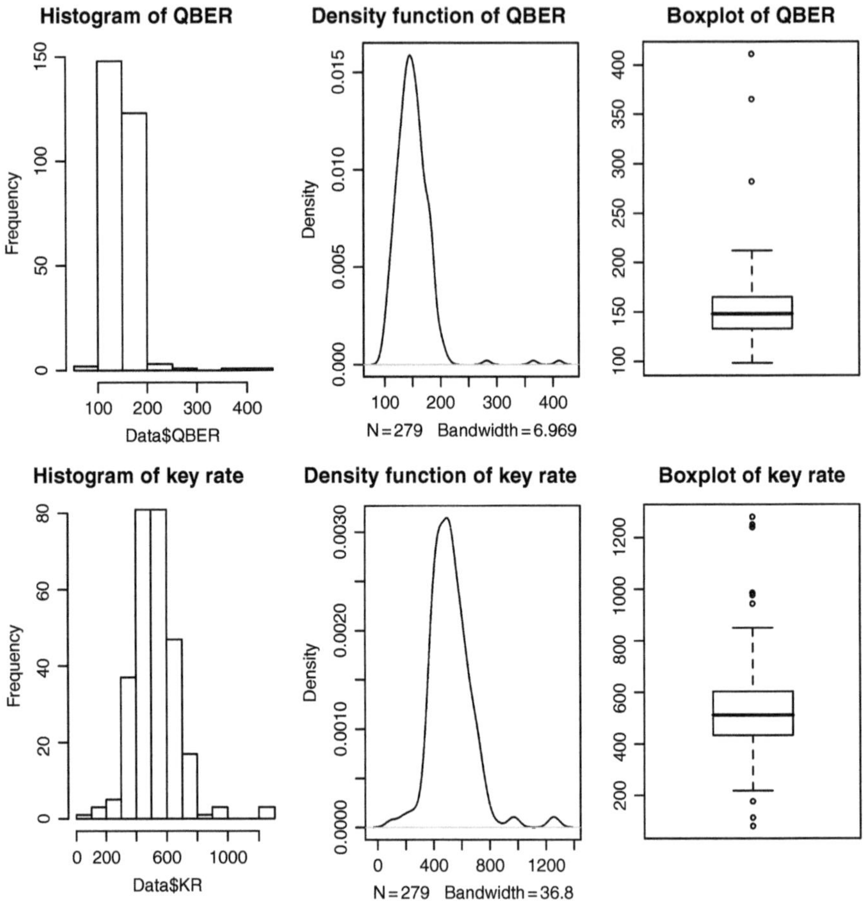

Fig. 7.6 Histogram, density function, and boxplot of QBER and KR of the device "Autocompensating Plug&Play"

Table 7.11 lists the summary statistics, i.e., minimum, 1st quantile, median, mean, 3rd quantile, and maximum for the device "Continuous Variables".

In Fig. 7.7 the scatterplot matrix for the data set of the device "Continuous Variables" can be seen. A high correlation between QBER and KR is obvious. The

Table 7.10 Extract of the data set for the device "Continuous Variables"

QBER	KR	Temperature	Humidity	Sunshine duration	Global radiation
272	6070	148	88	0	0
244	9039	145	89	0	0
331	9325	143	90	0	0
372	6897	141	91	0	0
286	9952	140	91	0	0
456	9686	141	91	0	0

Table 7.11 Numerical overview for the device "Continuous Variables"

QBER		KR		Temperature	
Min. :	17.0	Min. :	1223	Min. :	117.0
1st Qu.:	220.2	1st Qu.:	8045	1st Qu.:	135.0
Median :	327.5	Median :	9091	Median :	148.0
Mean:	407.6	Mean:	8620	Mean:	149.7
3rd Qu.:	586.2	3rd Qu.:	9764	3rd Qu.:	163.0
Max. :	1222.0	Max. :	12943	Max. :	184.0
Humidity		**Sunshine duration**		**Global radiation**	
Min. :	71.00	Min. :	0.0	Min. :	0.0
1st Qu.:	80.00	1st Qu.:	0.0	1st Qu.:	0.0
Median:	84.00	Median :	0.0	Median :	0.0
Mean :	84.05	Mean :	92.57	Mean :	89.39
3rd Qu.:	91.00	3rd Qu.:	0.0	3rd Qu.:	131.25
Max. :	93.00	Max. :	600.0	Max. :	539.00

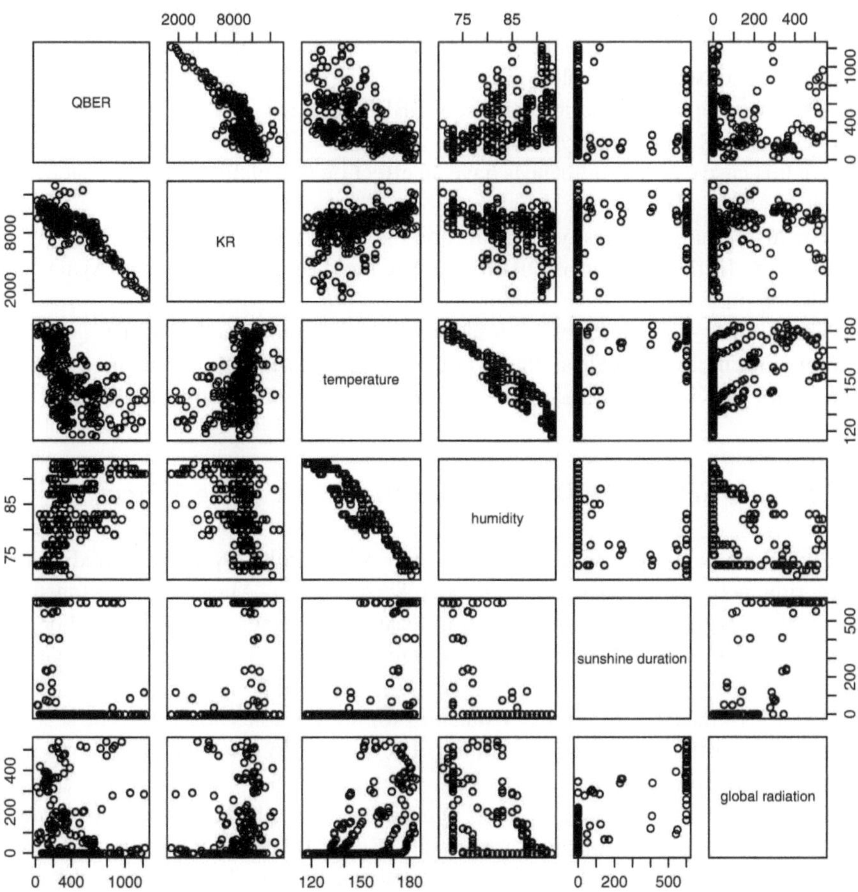

Fig. 7.7 Scatterplot matrix for the data set of the device "Continuous Variables"

Table 7.12 Correlation matrix for the device "Continuous Variables"

	QBER	KR	Temperature
QBER	1.0000000	−0.8567638	−0.5269060
KR	−0.8567639	1.0000000	0.4354725
Temperature	−0.5269060	0.4354724	1.0000000
Humidity	0.4427780	−0.3793859	−0.9480146
Sunshine duration	−0.1716637	0.1229877	0.5661680
Global radiation	−0.1167913	0.0804763	0.5411758

	Humidity	Sunshine duration	Global radiation
QBER	0.4427780	−0.1716637	−0.1167913
KR	−0.3793859	0.1229877	0.0804763
Temperature	−0.9480146	0.5661680	0.5411757
Humidity	1.0000000	−0.6013030	−0.5501430
Sunshine duration	−0.6013030	1.0000000	0.8546850
Global radiation	−0.5501431	0.8546851	1.0000000

correlation matrix – Table 7.12 – gives the numerical values for the correlations between the observed variables.

There is a high correlation between QBER and KR and a small correlation between QBER and temperature and also between QBER and humidity. It seems that in this case external influences have an effect on the quality of QKD systems of the device type "Continuous Variables."

The histogram, the density function, and the boxplot of QBER and KR are given in Fig. 7.8. In the boxplot of QBER and KR the outliers can be seen. Again, for prediction purposes, we have eliminated the outliers.

The Cramer–von Mises two sample test supports the hypothesis that QBER is gamma distributed as well as the hypothesis that it is log-normal distributed. As before, for the GLMM in Sect. 7.3.2, QBER is taken to be gamma distributed.

7.2.5 Data Set for the Device "One Way Weak Pulse System"

The measurements for the device "One Way Weak Pulse System" started in the fiber ring network in Vienna on October 8, 2008, too. Alice was positioned in Breitenfurterstrasse (BRT) and Bob was located in Siemensstrasse (SIE). For the measurements the route BRT – SIE was used. In Table 7.13 a short extract can be seen.

Table 7.14 lists the six summary statistics for the variables observed. The scatterplot matrix for the data set is given in Fig. 7.9. Table 7.15 gives the exact numerical values for the correlations.

In the correlation matrix a slight correlation between QBER and KR can be seen. On the other hand, a significant correlation between QBER and temperature is obvious. There is also a correlation between QBER and humidity, sunshine duration and global radiation. Due to this it can be assumed that external influences have an effect on the quality of QKD systems of the device type "One Way Weak Pulse Sys-

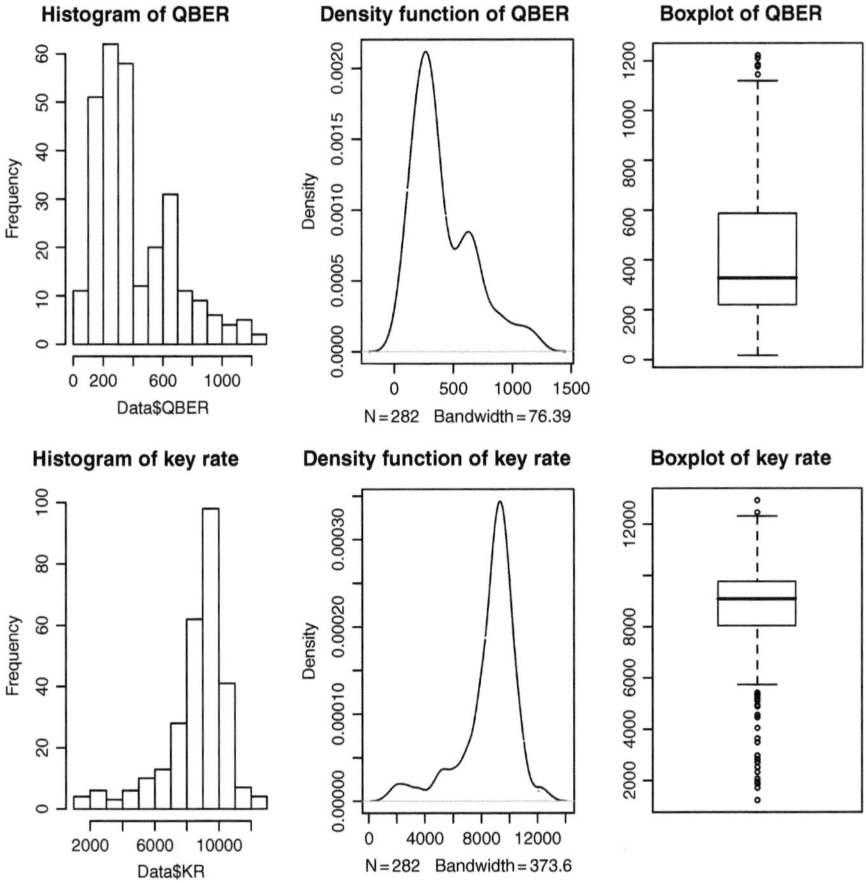

Fig. 7.8 Histogram, density function, and boxplot of QBER and KR of the device "Continuous Variables"

Table 7.13 Extract of the data set of the device "One Way Weak Pulse System"

QBER	KR	Temperature	Humidity	Sunshine duration	Global radiation
254	2463	161	81	0	0
250	2418	161	81	0	0
257	2448	160	82	0	0
283	3203	160	82	0	0
307	1987	158	83	0	0
254	2884	158	83	0	0

tem." The histograms, density plots, and boxplots of QBER and KR are displayed in Fig. 7.10.

The Cramer–von Mises two sample test declares that QBER is gamma distributed because the p-value is greater than 0.05. For the hypotheses of log-normal and normal distribution, respectively, the Cramer–von Mises two sample test statistics result in p-values smaller than 0.05; thus, these hypotheses are rejected.

Table 7.14 Numerical overview for the device "One Way Weak Pulse System"

QBER		KR		Temperature	
Min. :	224.0	Min. :	1	Min. :	127.0
1st Qu.:	248.0	1st Qu.:	3007	1st Qu.:	139.0
Median :	264.0	Median :	4002	Median :	148.0
Mean:	260.9	Mean:	3515	Mean:	149.3
3rd Qu.:	269.0	3rd Qu.:	4552	3rd Qu.:	155.0
Max. :	315.0	Max. :	5482	Max. :	184.0
Humidity		**Sunshine duration**		**Global radiation**	
Min. :	71.00	Min. :	0.0	Min. :	0.0
1st Qu.:	81.00	1st Qu.:	0.0	1st Qu.:	0.0
Median:	83.00	Median :	0.0	Median :	0.0
Mean :	83.61	Mean :	116.9	Mean :	111.4
3rd Qu.:	87.00	3rd Qu.:	0.0	3rd Qu.:	146.0
Max. :	93.00	Max. :	600.0	Max. :	539.00

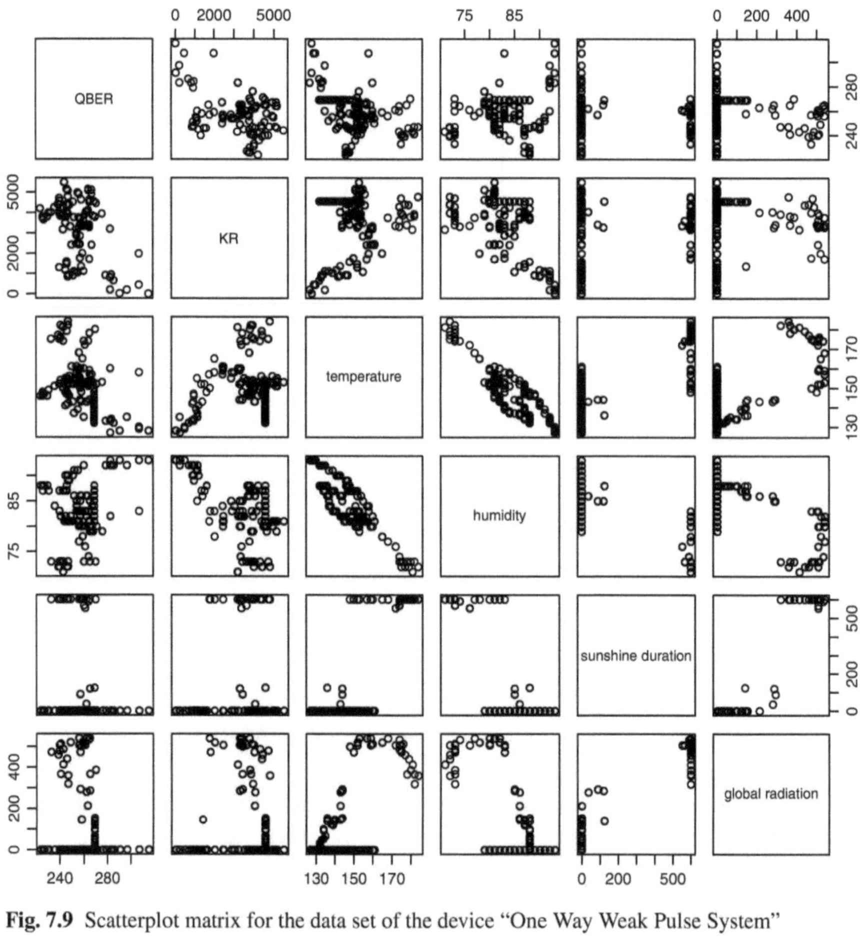

Fig. 7.9 Scatterplot matrix for the data set of the device "One Way Weak Pulse System"

Table 7.15 Correlation matrix for the device "One Way Weak Pulse System"

	QBER	KR	Temperature
QBER	1.0000000	−0.2591815	−0.5015706
KR	−0.2591815	1.0000000	0.1886785
temperature	−0.5015706	0.1886785	1.0000000
humidity	0.3460800	−0.4721942	−0.8909994
Sunshine duration	−0.2456862	−0.0034571	0.7155210
global radiation	−0.2023045	0.0234804	0.6072473
	Humidity	**Sunshine duration**	**Global radiation**
QBER	0.3460800	−0.2456861	−0.2023044
KR	−0.4721943	0.0034571	0.0234804
Temperature	−0.8909994	0.7155210	0.6072472
Humidity	1.0000000	−0.6756165	−0.5884132
Sunshine duration	−0.6756166	1.0000000	0.9519974
Global radiation	−0.5884132	0.9519974	1.0000000

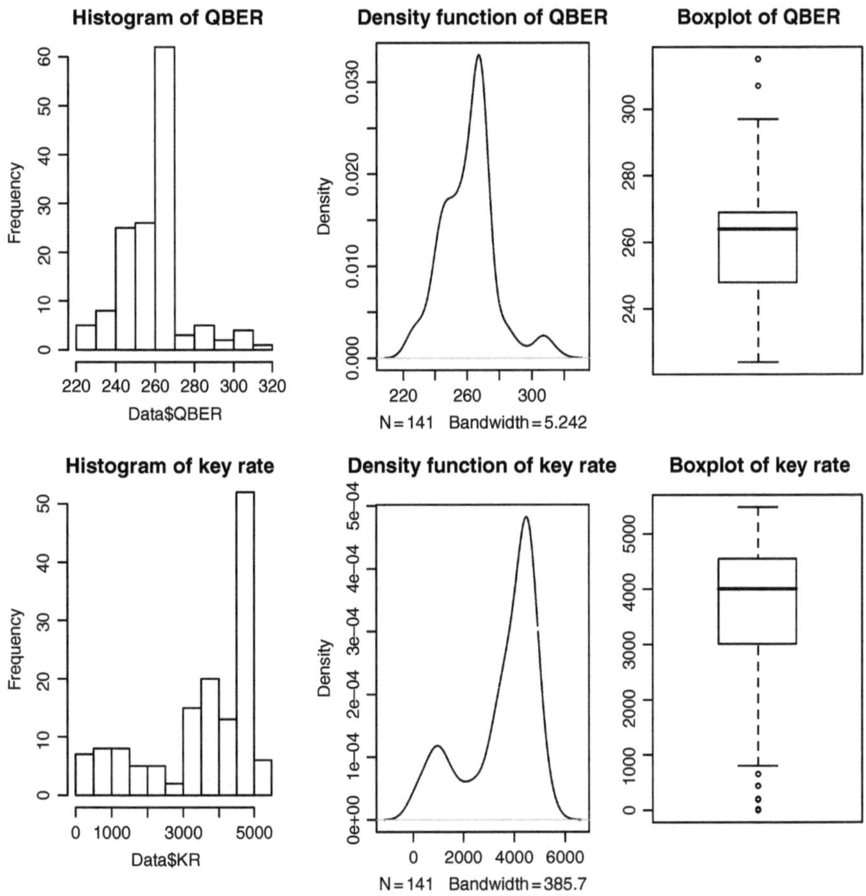

Fig. 7.10 Histogram, density function, and boxplot of QBER and KR of the device "One Way Weak Pulse System"

7.3 Statistical Analysis

The statistical analysis uses the generalized linear model introduced in Sect. 7.1.1 and furthermore the generalized linear mixed model introduced in Sect. 7.1.2. The implementations for the GLM and GLMM are based on [4] and [16].

7.3.1 Generalized Linear Model

For the illustration of the generalized linear model (GLM) only the data set of the entangled-based device (device "Entangled") is used. The Cramer–von Mises two sample test to the entangled-based data processed in Sect. 7.2.1 favors the gamma distribution of the data set. Due to this fact, the gamma generalized linear model is used which is intended for continuous, skewed responses. In this case the log link is used because it is consistent with the linear model. Subsequently, the gamma GLM with QBER as response, in dependency of the KR and external influences as predictors, is processed. In Table 7.16 the estimated regression coefficients $\hat{\beta}$ and the estimated error terms $\hat{\varepsilon}$ from the regression output can be seen.

Table 7.16 The estimated regression coefficients $\hat{\beta}$ and the estimated error terms $\hat{\varepsilon}$ from the GLM

Factor	$\hat{\beta}$	$\hat{\varepsilon}$	Unit
Intercept	6.452	$2.791 \cdot 10^{-1}$	1
KR	$-2.593 \cdot 10^{-4}$	$1.777 \cdot 10^{-5}$	bit/s
Temperature	$7.563 \cdot 10^{-4}$	$7.733 \cdot 10^{-4}$	°C/10
Humidity	$2.203 \cdot 10^{-4}$	$2.077 \cdot 10^{-3}$	%
Sunshine duration	$5.206 \cdot 10^{-5}$	$4.322 \cdot 10^{-5}$	s
Global radiation	$-2.514 \cdot 10^{-4}$	$5.695 \cdot 10^{-5}$	W/m^2

For demonstrating the goodness of fit of the current model several tests and criteria are used. One of these criteria is the Akaike Information Criterion (AIC). Usually, criteria for the goodness of fit of the regression model are based on the regression residuals. Moreover, leverage and influence can be used as criteria for the performance of a model. Concerning leverages it is important to know that they are somewhat different for GLMs. Other possibilities are to look at the jacknife residuals and the Cook statistics, respectively.

In contrast to the described residuals, leverage, and influence measures, also graphical methods such as a half-normal plot are used to detect points that do not fit to the model and consequently influence the fit unduly. In Fig. 7.11 a half-normal plot of the jacknife residuals is given. It seems that the data values 91 and 267 are outliers.

The half-normal plot can also be used for positive-valued diagnostics, i.e., for leverages and the Cook statistics. In Fig. 7.12 a half-normal plot of the leverages is shown. There is some indication that the data values 106 and 108 may have some leverage.

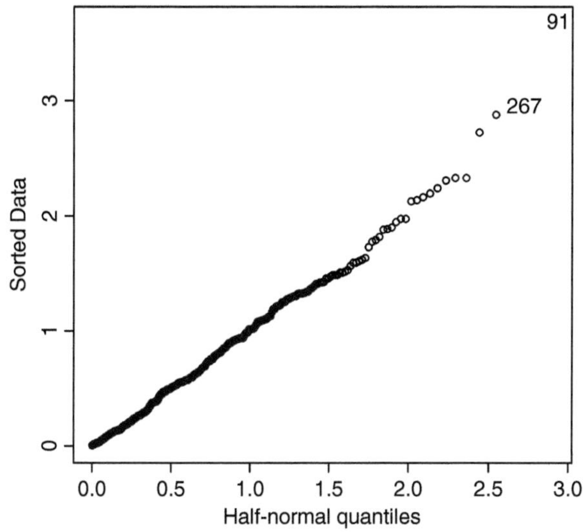

Fig. 7.11 Half-normal plot of the jacknife residuals

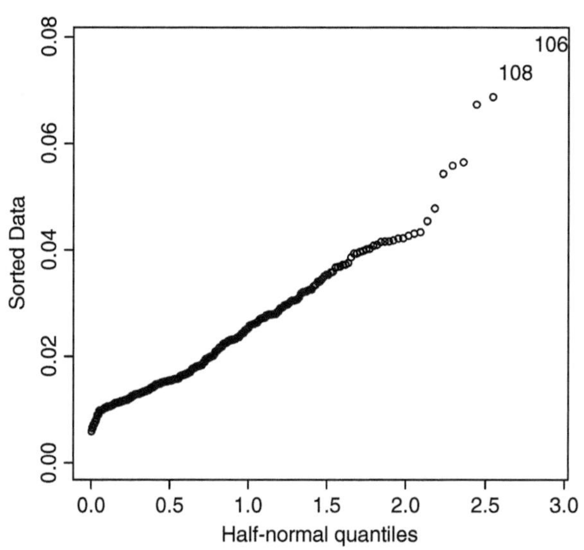

Fig. 7.12 Half-normal plot of the leverages

To confirm the results of the jacknife residuals and the leverages a half-normal plot of the Cook statistics is given in Fig. 7.13. It seems that the data values 78 and 91 have a higher influence compared to the rest of the data.

Due to this results it can be assumed that the data values 78, 91, 106, 108, and 267 are outliers. To these assumptions the concerned values are omitted and the gamma GLM analysis is repeated.

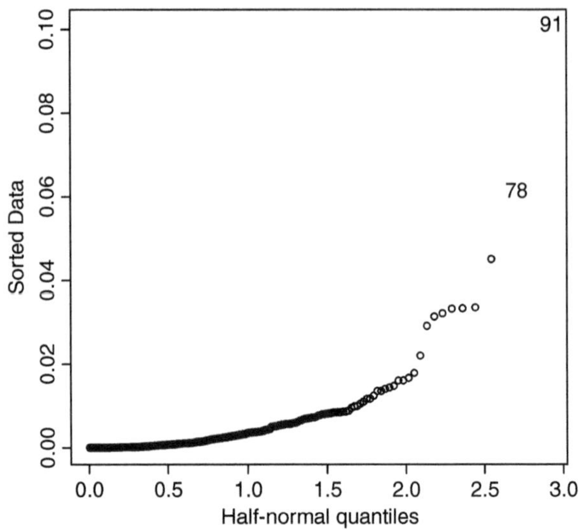

Fig. 7.13 Half-normal plot of the Cook statistics

Table 7.17 The estimated regression coefficients $\hat{\beta}$ from the original regression model and the estimated regression coefficients $\hat{\beta}_{wO}$ from the regression model without outliers

Factor	$\hat{\beta}$	$\hat{\beta}_{wO}$	Unit
Intercept	6.452	6.452	1
KR	$-2.593 \cdot 10^{-4}$	$-2.593 \cdot 10^{-4}$	bit/s
Temperature	$7.563 \cdot 10^{-4}$	$7.563 \cdot 10^{-4}$	°C/10
Humidity	$2.203 \cdot 10^{-4}$	$2.203 \cdot 10^{-4}$	%
Sunshine duration	$5.206 \cdot 10^{-5}$	$5.206 \cdot 10^{-5}$	s
Global radiation	$-2.514 \cdot 10^{-4}$	$-2.514 \cdot 10^{-4}$	W/m^2

From this regression output, the estimated regression coefficients $\hat{\beta}_{wO}$ and the regression coefficients $\hat{\beta}$ from the original regression output can be seen in Table 7.17.

No significant differences can be observed, which disagrees with the assumption that the data values 78, 91, 106, 108, and 267 are outliers. So, for the prediction the original regression model can be used.

In Sect. 7.2.1 the first six data points of the data set of October, 8, 2008 are shown. In the following the original regression model is used to predict QBER of these six data points. Comparing real and predicted QBER gives further information about the performance of the model. The comparison between real and predicted values can be seen in Table 7.18.

Moreover, in Table 7.19 an extract of the data set for the entangled-based device of October 11, 2008 is given. To predict QBER the original regression model with KR, temperature, humidity, sunshine duration, and global radiation as parameters is used. In Table 7.20 real and predicted QBER values are shown where a difference is obvious.

Table 7.18 Comparison between real and predicted QBER of GLM

Original	Predicted
329	355
345	362
329	355
336	317
379	377
354	325

Table 7.19 Extract of the data set for the entangled-based device of October 11, 2008

QBER	KR	Temperature	Humidity	Sunshine duration	Global radiation
351	2302	165	78	600	173
365	2244	165	78	600	146
358	2606	166	78	600	120
380	2498	165	78	600	93
365	2561	165	78	600	73
414	2041	165	78	339	47

Table 7.20 Comparison between real and predicted QBER of GLM

Original	Predicted
351	397
365	406
358	372
380	385
365	380
414	432

7.3.2 Generalized Linear Mixed Model

In Sect. 7.3.1 the GLM is used for prediction of the entangled-based device. Therefore, this regression model is useless for the prediction of other devices, i.e., for prediction of the device types "Freespace," "Autocompensating Plug&Play," "Continuous Variables," and "One Way Weak Pulse System." In the following the generalized linear mixed model (GLMM) is used to include this random effect caused by the different devices. The idea in this case is to involve the random effect by introducing a single random intercept for each device.

As mentioned in Sect. 7.1.2 the complex problem is to estimate the model parameters. Here a penalized quasi-likelihood approach is used.

In Table 7.21 the output of the GLMM is represented. There the fixed effects estimates, i.e., the estimated regression coefficients $\hat{\beta}$, and their approximated standard errors, i.e., the estimated error terms $\hat{\varepsilon}$, corresponding to the different fixed effects are shown. The random components of the model are also of interest. In Table 7.22 the values for the random intercepts of the different devices are given. In the random intercepts a clear difference between the devices is obvious, which confirms the assumption that there is a random effect caused by the five device types.

Table 7.21 The estimated regression coefficients $\hat{\beta}$ and the estimated error terms $\hat{\varepsilon}$ from the GLMM

Factor	$\hat{\beta}$	$\hat{\varepsilon}$	Unit
Intercept	$-8.474 \cdot 10^{-3}$	$1.746 \cdot 10^{-3}$	1
KR	$2.227 \cdot 10^{-7}$	$1.130 \cdot 10^{-8}$	bit/s
Temperature	$4.074 \cdot 10^{-5}$	$4.076 \cdot 10^{-6}$	°C/10
Humidity	$6.346 \cdot 10^{-5}$	$1.097 \cdot 10^{-5}$	%
Sunshine duration	$4.968 \cdot 10^{-7}$	$2.513 \cdot 10^{-7}$	s
Global radiation	$-1.261 \cdot 10^{-6}$	$3.118 \cdot 10^{-7}$	W/m^2

Table 7.22 The values for the random intercepts of the different devices from the GLMM

No.	Device	Intercept
1	Entangled	$-6.800 \cdot 10^{-4}$
2	Freespace	$-1.195 \cdot 10^{-3}$
3	Autocompensating Plug&Play	$3.736 \cdot 10^{-3}$
4	Continuous Variables	$-2.110 \cdot 10^{-3}$
5	One Way Weak Pulse System	$2.493 \cdot 10^{-4}$

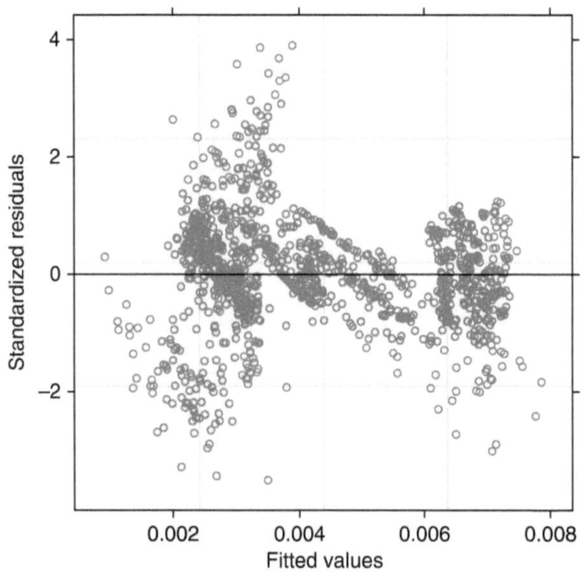

Fig. 7.14 Fitted values versus standardized residuals

Moreover, to get information about the goodness of fit of the current model the standardized residuals are considered. In Fig. 7.14 the fitted values versus the standardized residuals can be seen.

To increase information about the performance we try to predict the QBER. In Table 7.23 the real QBER of the data set of the entangled-based device of October 8, 2008 (processed in Sect. 7.2.1) is compared with the predicted QBER of the GLM

Table 7.23 Comparison between real QBER, predicted QBER (GLM), and predicted QBER (GLMM)

Real QBER	Predicted QBER (GLM)	Predicted QBER (GLMM)
329	355	325
345	362	334
329	355	334
336	317	325
379	377	347
354	325	328

and the predicted QBER of the GLMM. The prediction of the GLMM approximates the real QBER much better than the prediction based on the GLM.

A comparison between the real QBER and the predicted QBER of the GLMM is given in Table 7.24. There are also predictions for the other four devices "Freespace," "Autocompensating Plug&Play," "Continuous Variables," and "One Way Weak Pulse System." In the first column of Table 7.24, the data point of the data set can be seen which is used for prediction. In the second column the corresponding device is indicated. In the other two columns the real and the predicted QBER values are opposed. Generally, the predicted values based on the GLMM approximate the real values quite well.

7.4 Summary

In the statistical analysis, regression models such as generalized linear models (GLM) and generalized linear mixed models (GLMM) are used to ascertain whether external influences like temperature, humidity, sunshine duration, and global radiation effect the quality of QKD systems. Furthermore, these two regression models are, in consequence, used to predict the qubit error rate (QBER). The advantage of GLMs and GLMMs is that in both regression models it is possible to handle non-normal responses. Moreover, in a GLMM, random effects can be included.

A prototype of a QKD network was implemented in Vienna where the measurements of the different devices have been performed. The measurements started on October 1, 2008 and were concluded on November 8, 2008. The measurements for this chapter are based only on the data of October 8, 2008 to October 10, 2008 where the data sets of five different devices are used, i.e., of the device types "Entangled," "Freespace," "Autocompensating Plug&Play," "Continuous Variables," and "One Way Weak Pulse System." From these measurements the QBER and the key rate are available. From the Central Institute for Meteorology and Geodynamics (ZAMG) the data of external influences, i.e., temperature, humidity, sunshine duration, and global radiation, are obtained.

Statistical methods are used to process the data sets for the different devices. Furthermore, GLMs and GLMMs are used to get further information and to predict QBER. The conclusion that can be drawn from the statistical analysis is that temperature, humidity, sunshine duration, and global radiation have no influence

Table 7.24 Comparison between real and predicted QBER for different devices

Data point	Device	Real QBER	Predicted QBER
1	Entangled	329	325
2	Entangled	345	334
3	Entangled	329	334
4	Entangled	336	325
5	Entangled	379	347
6	Entangled	354	328
320	Freespace	209	223
321	Freespace	206	198
322	Freespace	205	163
323	Freespace	205	161
324	Freespace	208	190
325	Freespace	208	196
772	Autocompensating Plug&Play	117	151
773	Autocompensating Plug&Play	128	151
774	Autocompensating Plug&Play	122	152
775	Autocompensating Plug&Play	122	153
776	Autocompensating Plug&Play	141	153
777	Autocompensating Plug&Play	151	155
778	Continuous Variables	272	419
779	Continuous Variables	244	334
780	Continuous Variables	331	329
781	Continuous Variables	372	404
782	Continuous Variables	286	321
783	Continuous Variables	456	323
1183	One Way Weak Pulse System	284	304
1184	One Way Weak Pulse System	283	311
1185	One Way Weak Pulse System	279	309
1186	One Way Weak Pulse System	286	318
1187	One Way Weak Pulse System	291	345
1188	One Way Weak Pulse System	283	345

on the quality of the QKD devices "Entangled," "Freespace," and "Autocompensating Plug&Play." However, it could be detected that temperature, humidity, sunshine duration, and global radiation have a low influence on the quality of the QKD devices "Continuous Variables" and "One Way Weak Pulse System." For prediction purposes it is much better to use the GLMM instead of the GLM.

References

1. Anderson, T.W.: On the distribution of the two-sample Cramer-von mises criterion. Ann. Math. Stat. **33**(3), 1148–1159 (1962)
2. Demidenko, E.: Mixed Models. WILEY, Hoboken, NJ (2004)
3. Fahrmeir, L., Kneip, T., Lang, S.: Regression. Springer Verlag, Berlin (2007)
4. Faraway, J.J.: Extending the Linear Model with R. Chapman & Hall/CRC, Boca Raton, FL (2006)
5. Lessiak, K., Kollmitzer, C., Schauer, S., Pilz, J., Rass, S.: Statistical Analysis of QKD Networks in Real-life Environments. Third International Conference on Quantum, Nano and Micro Technologies (2009)

6. McCullagh, P., Nelder, J.: Generalized Linear Models, 2nd ed. Monographs on Statistics and Applied Probability 37. Chapman & Hall, London (1989)
7. Nelder, J.A., Wedderburn, R.: Generalized linear models. J. R. Stat. Soc. A **135**(3), 370–384 (1972)
8. Pinheiro, J.C., Bates, D.M.: Approximations to the log-likelihood function in the nonlinear mixed-efffects model. J. Comput. Graph. Stat. **4**(1), 12–35 (1995)
9. Pinheiro, J.C., Bates, D.M.: Mixed Effects Models in s and s-plus. Springer Verlag, New York (2000)
10. Poppe, A., Peev, M., Maurhart, O.: Outline of the SECOQC Quantum-Key-Distribution Network in Vienna. Int. J. of Quant. Inf. **6**(2), 209–218 (2008)
11. Sinha, S.K.: Robust analysis of generalized linear mixed models. J. Am. Stat. Assoc. **99**, 451–460 (2004)
12. Tierney, L., Kadane, J.B.: Accurate approximations for posterior moments and marginal densities. J. Am. Stat. Assoc. **81**, 82–86 (1986)
13. Verbeke, G., Molenberghs, G.: Linear Mixed Models for Longitudinal Data. Springer Verlag, New York (2000)
14. Verbeke, G., Molenberghs, G.: Models for Discrete Longitudinal Data. Springer Verlag, New York (2005)
15. Wolfinger, R., O'Connell, M.: Generalized linear mixed models: A pseudo-likelihood approach. J. Stat. Comput. Simul. **48**, 233–243 (1993)
16. Wood, S.N.: Generalized Additive Models. Chapman & Hall/CRC, Boca Ration, FL (2006)
17. Zeger, S.L., Liang, K.Y., Albert, P.S.: Models for longitudinal data: A generalized estimating equation approach. Biometrics **44**(4), 1049–1060 (1988)

Chapter 8
QKD Networks Based on Q3P

O. Maurhart

This chapter explains how single QKD links can be connected together via Q3P to form a QKD network. The term "network" emphasizes the need to interconnect more than two nodes. In this network arbitrary user data is transmitted from and to non-adjacent parties.

Here Q3P – the Quantum Point-to-Point Protocol – is described. This protocol expands the concepts of the well-known Point-to-Point Protocol (described in [12]) enabling and extending its prime ideas to adopt them in a quantum cryptography environment. Thus Q3P can guarantee information theoretical security between nodes of a QKD network with respect to encryption and authentication.

Based on Q3P, additional protocols for routing and message transmission are discussed which respond to QKD network differences in comparison to classical ones. Availability of key material, considerations of resend mechanisms, and trusted paths are among the properties which clearly separate QKD networks from other network types.

The protocols presented have been designed and developed during the EU founded SECOQC project and were successfully presented in Vienna, 2008 (see [9]).

8.1 QKD Networks

By means of QKD a shared secret key can be established on two sides, likely labeled Alice and Bob. Furthermore, quantum cryptography provides mechanisms and techniques to allow an eavesdropper (Eve) to be detected on the shared communication media.

The various techniques involved in QKD always use a quantum channel to transmit quantum information as well as a classical channel for further actions. In order to successfully establish a key between Alice and Bob to be used for one-time-pad

O. Maurhart (✉)
Safety & Security Department, Quantum Technologies, AIT Austrian Institute of Technology GmbH, Lakeside B01A, 9020 Klagenfurt, Austria, `oliver.maurhart@ait.ac.at`; `http://www.ait.ac.at`

Maurhart, O.: *QKD Networks Based on Q3P*. Lect. Notes Phys. **797**, 151–171 (2010)
DOI 10.1007/978-3-642-04831-9_8 © Springer-Verlag Berlin Heidelberg 2010

(OTP) and other cryptographic techniques of information theoretical security processing a number of steps have to be undertaken. Recalling previous chapters of this book to a bare minimum these are as follows:

- Sifting: This allows Alice and Bob to remove bits from their raw key due to wrong basis choice. After this step Alice and Bob share a *sifted* key. There have been various protocols developed in the recent years, starting with [1] and are still emerging since new quantum technologies are still discovered requiring new sifting protocols.
- Error Detection and Correction: Still there can be errors present in both keys. This is due to some quantum bits that may have been damaged or flipped during transmission or quantum detection failures. Note that errors are also probably introduced by an eavesdropper. Therefore the system has to decide whether the detected errors can reasonably be found by technological and/or physical shortcomings *or* are introduced due to the presence of an attacker and more importantly whether relevant security limits are exceeded. If the error rate can still be tolerated an error correction scheme is to be applied, otherwise the current key distillation round has to be terminated and the results skipped. The most popular error correction technique is CASCADE, but other more refined versions are researched and uprising. The result of this step is a *reconciled* key.
- Privacy Amplification: An eavesdropper – Eve – is always expected. As Eve is conservatively assumed to be the source of all errors occurring in the raw key acquisition phase and is supposed to listen on the insecure public channel on which the error correction steps above are performed, she may gather enough knowledge to deduce part of the key. Hence Alice and Bob now try to minimize Eve's knowledge by appropriately reducing the length of the reconciled key. After this step Alice and Bob have a *shared secret* key.

However, the benefit of QKD is restricted to a common shared quantum media. Therefore the key distillation steps described above are only applicable on a peer-to-peer basis, where Alice and Bob have access to the same quantum media. Also with respect to different quantum techniques key establishment may be expensive both in time and computational resources.

QKD networks now try to exceed point-to-point limitations, enabling Alice to send information to Bob without sharing a quantum channel. They are rather connected along a path of intermediate nodes. Therefore a QKD network is described by

- A quantum network consists of nodes connected by quantum channels. The quality and design of the quantum channels are limited by the certain technology used at and the physical attributes they impose. Therefore, quantum networks may be composed of quantum channels of different kind regarding distance and bandwidth compared to cost to establish and maintain.
- All nodes are connected to a classical communication media, which ought to be public and insecure. The nodes need this media for their specific QKD-protocols.

The issue, whether this media is the same or not as used for the quantum channel (i.e., fiber optic), is irrelevant from a conceptional point of view.

- A quantum channel *itself* is always a point-to-point link between exactly two nodes. One-to-many links are seen as multiple logical point-to-point links in spatial or time division manner. Routing protocols on top bypass the peer-to-peer limitations and enforce multi- and broadcast capabilities on the system.
- A quantum network node may participate in more than one, single quantum channel, thus making it a relay station.
- Messages between two adjacent nodes can be transferred with information theoretical security using at least one-time-pad cipher for encryption on the classical channel.
- Messages between two adjacent nodes can be authenticated with information theoretical security.

This results in delivering a message on a hop-by-hop basis as illustrated in Fig. 8.1. On receiving, each node decrypts the message using a local key shared with its neighbor and then retransmits it with a key it has for its other neighbor. The keys are discarded thereafter. Exhibiting such behavior qualifies QKD nodes to be viewed as "Store & Forward" network devices.

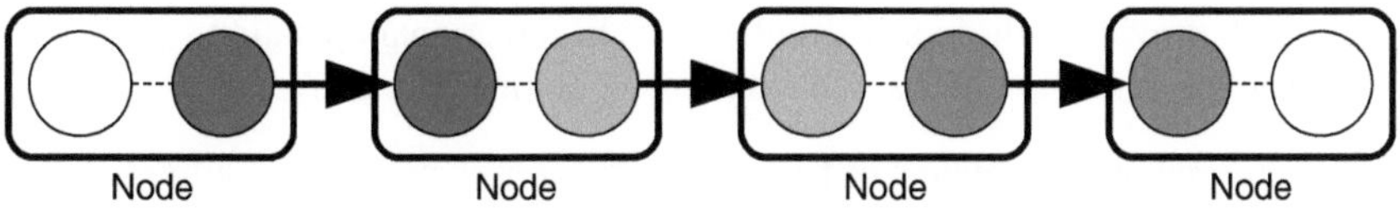

Fig. 8.1 Hop-by-hop transmission of a message between nodes

Having such a set of interconnected QKD nodes which transmit messages on a hop-by-hop basis constitutes to a *trusted repeater QKD network*. Topologies of such networks include mesh- and starlike networks. Within a trusted repeater QKD network, keys are shared between two parties along a path of intermediate and trustworthy nodes. This is opposed to quantum switching network in which the quantum channel is switched. The later model does connect more than two peers but is bound in distance. As QKD switched networks create selective point-to-point connections on demand by short-circuiting QKD channels, attenuation is seen as a major concern and problem. With the help of trusted repeaters any number of participating users connected across arbitrary distances can share secrets on a QKD network.

Obviously a trusted repeater QKD network can be built using more than one single QKD technology. The requirement to integrate different QKD techniques and methods into one single network interconnecting more than two users is indispensable to realize trusted repeater QKD networks. This reasons the need for a layer to be introduced which separates key usage from the technical details of key generation. This is much like today's Internet in which the client browser does not care about the concrete transport technique (Ethernet, WLAN, ...). This layering is solved by Q3P.

The prime building block of a quantum network is Q3P. As Q3P itself is deeply inspired by PPP, the classical, traditional protocol for interconnecting two machines, a brief summary of PPP has to be stated.

8.2 PPP

The Point-to-Point Protocol [12] standardized 1994 is used to establish a connection between two nodes in a network. It is primarily used as dial-up connection mechanism and superseded the former SLIP [11]. SLIP lacked error detection and was bound to serial lines as well to IP messaging solely.

PPP holds

- Encapsulation of various datagram protocols like IP, IPX, and AppleTalk
- Error Detection
- Link Control Mechanism
- Authentication

PPP does not rely on a certain connection technology as SLIP does (serial line). Instead PPP imposes a few requirements on the lower layer:

- The link provides full-duplex simultaneous bi-directional operation.
- All packets arrive in order at the responder side (in-order delivery).

PPP enforces no sequencing or fragmentation on packets which requires a collection and reordering facility at the responder side. PPP does not expect this options of the lower layer requiring to send information packed in one single PPP frame. However, information datagrams of higher layers can be distributed across several packets but it is the responsibility of the higher level protocol to reassemble the fragments orderly.

The feature list of PPP is as follows:

- Encapsulation: PPP is capable of encapsulating various protocols situated at the next higher layer when speaking in terms of the ISO/OSI model [13]. The list of encapsulated protocols includes IP, IPX, AppleTalk, and others. The mechanism provided in PPP is of a much general term, allowing any arbitrary protocol to be used within PPP.
- Link Control Protocol (LCP): a major part of PPP is dedicated to establish and configure the link.
- Network Control Protocol (NCP): to be independent of any succeeding network protocol running on top of PPP, a framework for negotiating several aspects and options of the higher protocol is presented. For each network protocol collaborating with PPP a specific NCP is written, i.e., IPCP – the Internet Protocol Control Protocol for IP [7].

PPP is used as a layer 2 protocol for the well-known ISO/OSI reference model but is also implemented on top of already existing layer 2 protocols like PPPoE [6]

– "Point-to-Point Protocol over Ethernet" –, PPtP [4] – "Point-to-Point Tunneling Protocol," – and PPPoA [3] – "Point-to-Point Protocol over ATM."

The intention of PPP is to be self-configuring and functionable right out-of-the box. Simplicity in maintenance and configuration and flexibility and expandability are key terms of PPP.

8.3 Q3P

Q3P is the *Quantum Point-to-Point Protocol*. Q3P uses the ideas in the well-known PPP to form a PPP-*like* protocol between two adjacent nodes in a quantum network.

The whole Q3P design serves different aspects:

- Provide a communication primitive between exactly two adjacent nodes in a quantum network to enable sending and receiving of messages enhanced by information theoretical security.
- Enforce overall security of a QKD node by being the very single but neat interface to the outside world on the classical channel.
- Utilize different and yet even unknown QKD protocols and devices for key generation.
- Utilize different and yet even unknown cryptographic algorithms for information theoretical security for encryption and authentication.
- Separate key production from key usage. Even more: leverage Quality of Service methods on key usage regardless of any concrete QKD machine producing them.

With respect to some ISO/OSI-like hierarchical structure, viewed "from top" Q3P offers the very same interface of level 2 regardless of any concrete quantum technique used underneath. This is required to build stable network protocols on top of it without the need to tackle with QKD implementation specifics and quirks.

8.3.1 Q3P Building Blocks

Figure 8.2 shows the very basic setup for a Q3P implementation. The setup shown consists of two QKD devices, a Q3P engine, a classical channel, and an application using Q3P facilities (information theoretical secure encryption and authentication to another direct connected peer). The central point is that both – the QKD devices and the application – use Q3P to communicate with their peers on the classical, public channel. The Q3P engine acts as a singleton, that is: there must be only one single instance in a system running.

The sketch shows two instances of QKD devices, each of them having their own dedicated quantum channel. It is possible to connect more than one QKD device to a single Q3P engine. Having more than one QKD device connected to a Q3P engine opens up multiple connections to different peers, the basic building block for a network.

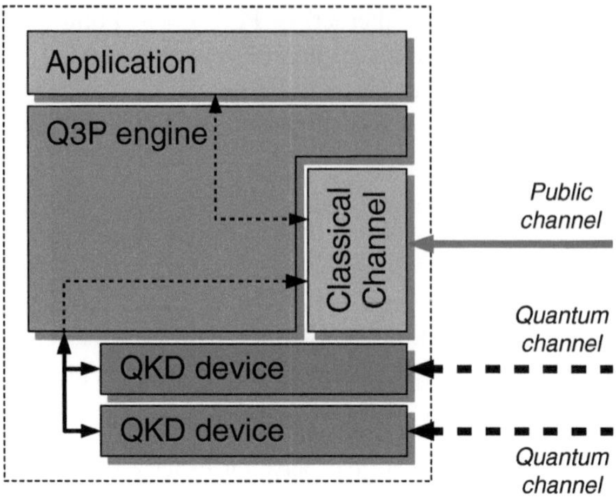

Fig. 8.2 The "Q3P Engine" where all Q3P processing takes place

Examining the Q3P engine further we see:

1. The *Q3P node*. This entity represents administration and coordination unit. It creates Q3P link entities on demand. Whenever a connection from the outside – be it from a QKD device, a peer Q3P node or any application utilizing Q3P – is made the Q3P node is the very first to be contacted. It dispatches requests and commands to the proper Q3P link processes. As the node is responsible for creating the Q3P link instances it also observes, terminates, and relaunches them as appropriate.
2. The *Q3P link*. A Q3P link is a concrete instance of an established connection between two peers. The Q3P link entity is responsible for encryption, authentication, and authentication check. A single Q3P link instance serves all QKD to QKD connections between two nodes. That is, on Q3P node *A* and node *B*, we attach an arbitrary number of QKD devices and interconnect them each: still there is only one single Q3P link instance serving all QKD to QKD connections. Any messages which are sent and received by Q3P pass a Q3P link to get encrypted and authenticated or authentication checked and decrypted, respectively. As all messages are passed through this instance a Q3P link maintains *Channels* for each application to application (or QKD device to QKD device) communication line (Fig. 8.3).
3. The *Key Store*. Each Q3P link has an associated Key Store which buffers available key material for future use. Key Stores receive their key material from the QKD devices underneath. However, once a key has been delivered to such Key Stores any tags of information, stating the origin of the key material or the concrete date and time when it has been created, are removed. As such, a quantum key is nothing but a bitstream with quantum cryptography-based randomness.
4. Crypto Engine. The Crypto Engine acts as a container for any security – or more precisely cryptographic – relevant algorithms. Whereas OTP is rather

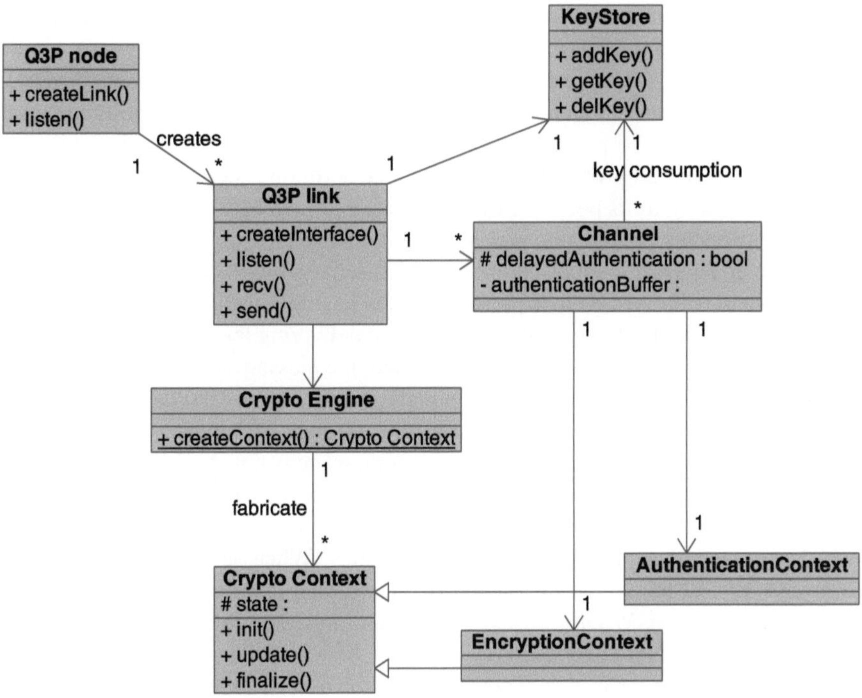

Fig. 8.3 Q3P class diagram in UML

trivial, information theoretical authentication is not. The Crypto Engine is capable of running different authentications with different tag sizes in parallel. This is achieved by creating *Crypto Contexts* dedicated either to the realm of encryption or authentication. Each channel has access to its unique encryption and authentication context.

8.3.2 Message-Based Streaming

Q3P connects two adjacent nodes. Such communication line which is hold up by the Q3P link entities on both sides is called a *Q3P Association*. A single Q3P Association serves more than one single application or QKD device. Each pair of QKD devices or applications utilizing a Q3P Association maintain a set of *Q3P Connections*. That is to say: a single Q3P Connection interconnects two QKD devices (or two applications, respectively). As QKD devices and applications may wish to send and receive in parallel each Q3P Connection can be further divided into *Q3P Channels*. It is these Q3P Channels to which cryptographic routines are attached and operated on.

Q3P is a message-stream based protocol. A single unit of communication for Q3P is a message. This is different for well-established protocols like TCP (see

[10]) for they are byte-streamed. Byte-streamed based protocols work on streams of bytes without interpretation of these streams in any way. On the other side, applications do have special semantics and meanings introduced into such byte-streams. Hence, even a video-stream encoded as MPEG or a VoIP-stream is split into several "messages." In order to preserve the message semantics each application utilizing a byte-streamed based protocol has to undertake additional steps to signal message starting and ending to its peer. For message-based protocols this is not needed. And for Q3P this comes natural since for a security centric protocol the key element for encryption and authentication schemes widely used is a message, not a byte-stream.

A message is defined as a sequence of bytes having a well-defined starting and ending position within the transport mechanism delivering the information. Further on – for Q3P – a message is not been interpreted; the message's content is of no concern to Q3P. However, messages are uniquely identifiable over a reasonable period of time. Ideally, there is no number clash of a message's ID during the lifetime of an established Q3P Connection. The size of a message is not limited and may vary, but performance considerations and hardware do impose restrictions.

The disadvantage of using a message-based view resides in the "store and forward" principle of nearly all networking facilities. When sending a message the receiver waits till the whole message arrived *and then* dispatches it for next computation (or routing). When messages are rather large then the transmission delay, the timespan encountered between receiving the very first byte of the message to the very last one, is as large as well. Thus enlarging the overall delay in communication.

However, knowing this, applications have the capability to increase (and reduce) performance on their own, by splitting their information into smaller messages and even drop some if their content does not provide essential meaning but simply increases quality of the data perceived. This is definitively true for media streams like MPEG which may drop B- and P-Frames off the whole message stream.

As Q3P has no knowledge of the concrete content of a message, it cannot decide which message can be dropped or not. Leaving this capability to entities having the knowledge of the concrete content of a message liberates Q3P to marginally rebuild parts of higher level protocols (Fig. 8.4).

8.3.3 Security Modes

The main goal of Q3P is to enforce a transmission of any messages by means of information theoretical security between two nodes. Q3P offers to send messages with these schemes:

- encrypted and authenticated
- authenticated
- plain

This is very contrary to most established security-focused network protocols like IPSec (IPSec starting with [5] and ongoing). Common security architectures within

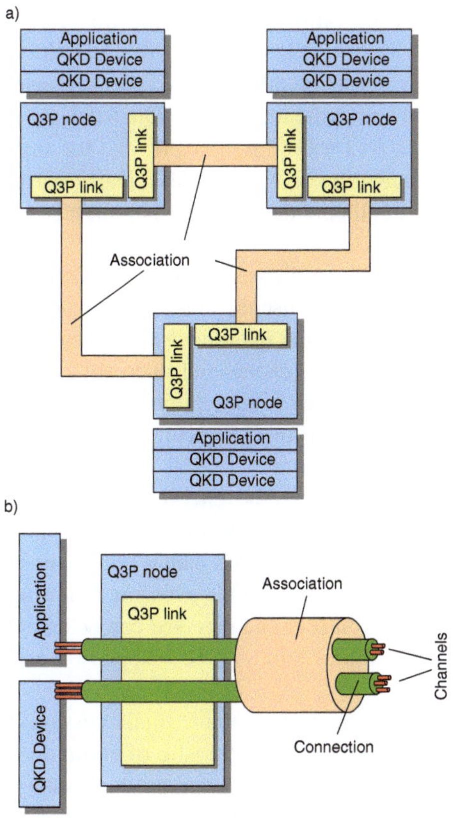

Fig. 8.4 Relationship between Q3P Associations, Q3P Connections, and Q3P Channels. (**a**) shows three Q3P nodes interconnected by their Q3P link instances via Q3P Associations. On each Q3P node an application and two QKD devices are attached communicating with their peers. (**b**) shows the situation on one single Q3P node: an application has a connection established. Within this Q3P Connection it utilizes two Channels. On the very same Association a QKD device has another Connection but with three Channels up and running

the networking community usually introduce an *all-or-nothing* principle whereas either the whole communication is encrypted and authenticated or no encryption and authentication is applied at all.

Secure transmission within Q3P is done by information theoretical security. As for the time of writing no *non*-key consuming algorithm with information theoretical secure quality is known, be it for encryption or be it for authentication. Therefore, any key used for encryption and/or authentication when sending messages is wasted if the message content is not of sensible nature for secrecy or integrity.

This is not a problem if QKD devices do produce much more keys than needed by applications. But today quantum key material is a very scarce resource and has to be used very sensitive. As bandwidth throughput is limited by the available key material it is counterproductive to waste such keys on messages not sensitive enough to be encrypted. This is the rationale in offering this variety of communication schemes.

Another reason is this: let M be the message. The keys used if the message is to be transmitted encrypted and authenticated are donated as k_E for the encryption and k_A for the authentication. As for OTP $|k_E| = |M|$ and for any arbitrary information theoretical secure authentication $|k_A| > 0$ holds, clearly stating $|M| < |k_E| + |k_A|$; with full encryption and authentication turned on, slightly more key bits are consumed than transmitted.

As mentioned earlier Q3P acts also as the only interface for the QKD devices on the classical channel and the latter use the classical channel to do their sifting, error correction, and privacy amplification to produce a new key. Therefore, the sum of the key material used for all messages exchanged to create a new key may now exceed the length of the new key produced if encryption and authentication is activated for each and every message. These results in having more key material consumed than created by the generation process.

To avoid this effect Q3P offers the encryption and authentication combined modes, authentication without encryption mode, and the plain mode without any security attached. The latter two are reasonable whenever information has to be exchanged of non-confidential nature. One even may omit authentication at all, if the integrity of the information sent and received can be determined in future steps, e.g., the application's communication can be mapped on a transaction processing concept.[1]

Another concern is authentication for many small messages. A huge load of small messages to be authenticated is a characteristic of common QKD error correction protocols like CASCADE. As QKD devices produce such messages and due to the nature of some error correction algorithms lacking possibilities to accumulate such messages and send them in once, another authentication scheme is introduced: *Delayed Authentication*. This authentication technique is offered along the *Instant Authentication* in which each and every message is authenticated at once.

Within Delayed Authentication (see Fig. 8.5), messages to be sent are locally copied to an internal Q3P buffer and then transmitted across the network to the peer. This particular transmission is *not* authenticated but sent in plain instead, consuming no key material at all. Gradually, the Q3P authentication buffer is filled with a series of small messages as the QKD device communicates with its peer.

Finally, the device requests an authentication-buffer check from the Q3P implementation at a suitable position within its processing QKD protocol. Q3P now computes the authentication tag and verifies this tag with the one received by the peer's Q3P. The result of this verification is responded back to the invoking QKD device. If authentication succeeded the device can be sure that all buffered messages so far were not altered by an adversary.

---

[1] A transaction concept having the ACID – Atomicity, Consistency, Isolation, and Durability – properties. A transaction has a clearly defined starting point and a clearly defined endpoint. Any action undertaken between these two points does not have any consequence to the surrounding environment until the full transaction is committed at the endpoint.

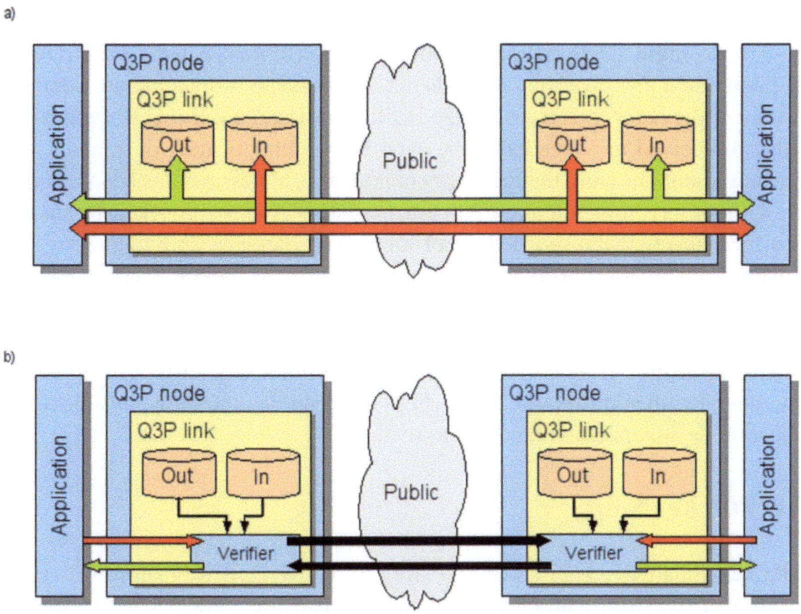

Fig. 8.5 Delayed Authentication. (**a**) shows the situation during normal message processing: all messages are copied to the outbound buffer before sending. On receiving the message is copied to the inbound buffer before handing out to the application (or QKD device). (**b**) illustrates the authentication check: both applications *have to* issue an authentication check. This stops ordinary message processing until the peer also ordered an authentication check as well. The authentication check is then performed by the Q3P Link instance (Verifier) and the Q3P Channel is unblocked on success again

Note that Q3P maintains a send and receive buffer for the delayed authentication separately. When performing delayed authentication now the authentication check is done for both directions, ingoing and outgoing at once. On receiving the inbound message buffer tag from the peer in this process the local party can also decide if information sent by it has been received untainted. This effect is quite uncommon in today's authentication protocols: giving the *sender* the ability to prove the authentication of the messages sent as well as the receiver. Though it seems that this design is memory consuming, authentication algorithms usually act on message blocks modifying some intermediate state S. When finalizing the authentication process some additional computation is done on the intermediate state S. This hints the way for memory optimization since for authentication algorithms which can be laid out this way only S is of concern and the message blocks once computed can be discarded for the authentication afterward.

Clearly, Q3P cannot state when such authentication checks have to be done. This is up to the application for only it knows the concrete context and find a proper position along its own, possibly proprietary application protocol. When doing OLTP (Online Transaction Processing) such steps ought to be found right before or along the COMMIT instruction.

Having the freedom of choice for a variety of several different security modes to choose for a single Q3P Channel introduces responsibility as well. If Q3P receives an authenticated message then it hands out the message if the authentication check succeeds. If the latter fails the message is silently discarded. But Q3P cannot decide whether a received message has to be encrypted, authenticated, or none of both in advance without knowing the message context, e.g., a message has been received in plain without any authentication tag attached. This *must be* checked by the application by inspecting the mode flags of a received message and comparing this to the internal application protocol: Is an unauthenticated message acceptable at the current stage of processing? Q3P has no means to tell if an unauthenticated message should be received authenticated – Q3P lacks the domain knowledge of the concrete application protocol to decide. This is a drawback since now applications are burdened with additional security checks introducing more responsibility on the application designers when it comes to security.

8.3.4 The Key Store

At the very heart of Q3P is the Q3P Key Store. A Q3P Key Store differs from any classical key storing and management infrastructure: once a key has been used, it is ought to be discarded. This is certainly true for OTP encryption and for any information theoretical authentication scheme currently used in Q3P.

These arguments request a total different approach than common public key systems or common symmetrical key buffers do imply – because they usually do not trash the key at first usage.

Four primitive functions have to be fulfilled:

1. Store a key.
2. Shred a key. The key store removes a key qualified by a given key identifier. This is natural if the key has been used. The notion "to shred" indicates that the key really has to be destroyed instead of simply "deleting" or "dropping" a key. Shredding a key may find its limitations by operating system memory management and by physical attributes of the memory media.
3. Retrieving a key. The key store offers a method to retrieve the *next* key with a given length l. A second technique is to retrieve a key in the database based on a key identifier.
4. Query the state of the key store. The key store has to give knowledge about various internal states to enforce QoS. That is, the amount of key material within the key store, statistics about recent key usage, and reasonable assumptions about key material growth.

Keys are delivered by QKD devices. However, they may do that in any variation of order. Next, the key material is used in symmetric key cryptographic algorithms and thus is not prequalified for any particular direction. When a party decides to send something to its peer it takes *any* key. There is no direction of data flow attached to a key nor any specific task label, like mutual dedication for encryption or

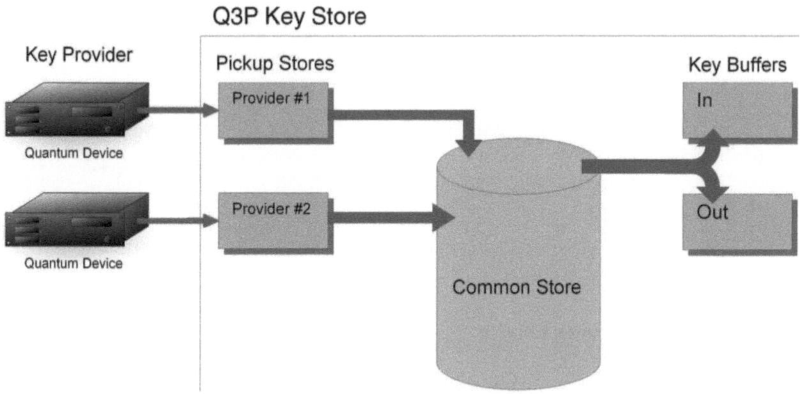

Fig. 8.6 Q3P Key store buffers: Pickup-stores, common store and in-out-buffers

authentication. To send and receive messages *some* keys are used. As the peer proceeds likewise clashes in key access are obvious leading to collisions when occasionally both parties decide to send data utilizing the very same key.

To solve this problem, three strategies are subject of concern: (a) try and detect collisions and, in case, recover, (b) use dedicated send and receive key material buffers on key delivery of a QKD device, or (c) authorization of key handout. Following the last two strategies the Q3P Key Store addresses these issues by introducing a *Master–Slave* relationship on multi-level key buffer.

Figure 8.6 depicts the several key buffers used in Q3P:

1. The key material pushed by the QKD devices is collected in the *Pickup Stores*. Here the keys may appear in any order. Q3P periodically picks up key material and verifies its existence with the peer.
2. Once it is assured that key material is present in both stores the keys are transferred to the *Common Store*. The Common Store represents a big mass of homogeneous key bits organized in blocks. Whereas keys found in the Pickup Stores can be associated with a certain QKD device and thus may even have some certain properties attached, like the error rate when producing this key, time stamp and others, keys in the Common Store don't have this additional information: key identifiers, time stamp, and any other attributes are detached. New keys transferred to this store are merged with already existing key bits there. Though key material is accessible by its BlockID at chunks of varying sizes.
3. The *In-Out-Buffers* hold key material dedicated for receiving and sending. Both buffers are symmetrically interlocked with the peer's buffers. That is, the In Buffer is synchronized with the peer's Out-Buffer and vice versa (see Fig. 8.7).

As opposed to the Pickup Store and the In-Out-Buffers the Common Store is the only key buffer which is persistent. Both other key storage devices are transient, meaning they are emptied at the start and at the end of a Q3P instance lifetime. Any key material placed in the Pickup Stores or the In-/Out-Buffers is lost if Q3P terminates. The Common Store places its key material on some media (e.g., Secured

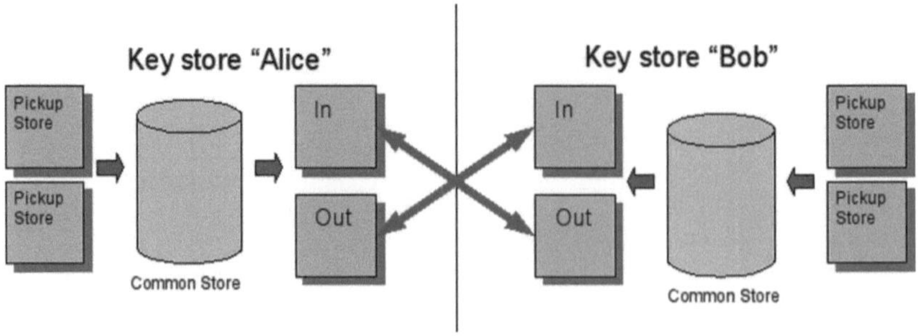

Fig. 8.7 Interleaved Q3P In-Out-buffers

Hard Disk, Smart Card, ...). On starting the Common Store is reincarnated with already collected and synchronized key material. This behavior is to some extent crucial since authentication protocols depend on preshared secrets, which can be now served from the Common Store.

In order to keep these buffers in synchronization, Q3P introduces a Master–Slave relationship. One of the Key Stores in a Q3P Association is the Master, the second the Slave. The central point thereby is that only the Master authorize key material to be transferred from a key buffer to another one.

Among the protocols which are needed for the Key Store to function properly the two most important subprotocols are described as follow. In order to transfer Key Material from the Pickup Store to the Common Store the Master Key Store ignites a STORE subprotocol.

```
Master                                                          Slave
   |                                                              |
   | MsgId-1, "STORE",                                            |
   |      (PickupStoreID, KeyID, BlockID)+, AUTH                  |
   |------------------------------------------------------------> |
   |                                                              |
   |       MsgId-2, "STORED", MsgId-1,                            |
   |                (PickupStoreID, KeyID, BlockID)*, AUTH        |
   |    <---------------------------------------------------------|
   |                                                              |
   | MsgId-3, "ACKNOWLEDGE", MsgId-2, AUTH                        |
   |----------------------------------------------------------->  |
   |                                                              |
```

Q3P key Store 3-phase STORE subprotocol

The data sent from the Master to the Slave is a MsgId-1, holding a unique number for this message, a STORE command identifier, followed by a minimum of 1 (PickupStoreID, KeyID, BlockID) tuple addressing the Pickup Store, the key within that Pickup Store and the BlockID to be assigned within the Common Store. The message itself is authenticated with an AUTH tag.

On response the Slave issues another MsgId-2, the number for the Slave's response message, a STORED keyword, MsgId-1 pointing to the former message of the Master for which this message is a response to, a series of (PickupStoreID, KeyID, BlockID) tuples – which may be empty – and finalized with another AUTH tag.

Finally, the Master responses in the same manner with an ACKNOWLEDGE keyword adding the message id of the Slave to which is an acknowledgment.

As the communication may break at any time the keys are not directly pushed to the Common Store at once. At step 1 – the Master sending the STORE command – the Master Key Store holds the series of keys taken from the Pickup Stores in suspense and does not place them in the Common Store now. On reception of the STORE command, the Slave selects the keys he has and creates the answer STORED command. However, the Slave pushes the keys to the Common Store but labels them as *potentially in sync* with the Master's Common Store. If the Master receives now the STORED answer from the Slave, the Master knows that the Slave has withdrawn these keys from the Pickup Store and therefore are present in the Slave's Common Store. The Master now writes the keys to the Common Store and labels them as *real in sync*. As the Slave gets the ACKNOWLEDGE message from the Master, he now relabels these keys in the Common Store as *real in sync*.

After the STORE subprotocol Master and Slave do have several keys which have been pushed by the QKD devices in sync. But in order to utilize these keys they have to be dedicated for a particular direction: inbound or outbound. Keys found in the Out-Buffer are dedicated to be used for sending messages. Therefore, on sending a key of suitable length is withdrawn from the Out-Buffer for encryption and/or authentication. Contrary, the In-Buffer holds keys which are used for decryption and authentication checks of received messages.

In order to fairly fill these two buffers, Master and Slave observe their In-Buffer. If the Master detects a shortage of key material in its In-Buffer it selects key material from the Common Store which is labeled as *real in sync* and issues the Slave to load this key material. This is done by the LOAD subprotocol. If the Slave encounters shortage of its In-Buffer it sends the Master a request to initiate a LOAD subprotocol but with different target Buffers. With a single LOAD command as many key material as possible is to be loaded but not more than $\lfloor |CommonStore|/2 \rfloor$. By having $d = |InBuffer| - |OutBuffer|$ "shortage" is present if $d \leqq 0$ on Master side and $d < 0$ on Slave side. The In-/Out-Buffers do have a maximum capacity, which may not be exceeded, since every key material present in these buffers is lost if the system goes down. As the encryption key is taken from the Out-Buffer the maximum length of a single message is defined by this limit.

```
Master                                                  Slave
   |                                                       |
   |                MsgId-1, "LOAD-REQUEST", Count, AUTH   |
   |   <---------------------------------------------------|
   |                                                       |
   | MsgId-2, "LOAD", Buffer, StartIndex,                  |
   | BlockId+, AUTH                                        |
   |------------------------------------------------->     |
   |                                                       |
   |               MsgId-3, "ACKNOWLEDGE", MsgId-2, AUTH   |
   |   <---------------------------------------------------|
   |                                                       |
```

Q3P-Protocol: 3-phase LOAD subprotocol with Slave's initiative

It is important to note that only the Master decides which key material is to be distributed to which Buffer. Since the Master chooses only *real in sync* key material he knows that this material must be present at the Slave's side; the same is not true for *potentially in sync* keys on the Master side. Also by observing the In-Buffer the communication paradigm is receiver-driven: the sender may not send more information than the receiver can compute; the receiver dictates the speed of transmission.

On startup the operator of a Q3P link decides which side becomes a Master and which one the Slave. As preshared secrets are needed for the first authentication by convention the Master picks up the first block of the Common Store and loads it into its Out-Buffer and the second into its In-Buffer. The Slave does likewise vice versa.

8.3.5 Q3P Packet Layout

A Q3P packet is defined as:

```
                1   1   2   2   2   3
0   4   8   2   6   0   4   8   2
+---+---+---+---+---+---+---+---+
|             Length            |
+---+---+---+---+---+---+---+---+
|             Msg-Nr            |
+---+---+---+---+---+---+---+---+
| Flags |Command|    Channel    |
+---+---+---+---+---+---+---+---+
|                               |
|              ...              |
|            Payload            |
|              ...              |
|                               |
+---+---+---+---+---+---+---+---+
|                               |
|            AUTH-Tag           |
|                               |
+---+---+---+---+---+---+---+---+
|             Key-ID            |
+---+---+---+---+---+---+---+---+
```

Q3P packet layout

At first the length of the packet is given, including the total header and trailer, followed by the message number. Next a flag indicates if the message has been encrypted and/or authenticated. The channel field holds the channel identifier. Then we find the payload. If the encryption flag is set the payload is the cipher else it is the plaintext. Presence of the authentication flag in the flags field requires the authentication tag to follow the payload part. Finally, the key identifier is added. This key identifier points into the In- or corresponding Out-Buffer as an offset. As encryption and authentication are always performed sequentially the address of one continuous stream of keys in the buffer is sufficient.

Hence that the authentication tag may have varying sizes. The concrete authentication algorithm to use and therefore the generated tag size must be set already by the operator out-of-band on a per Association basis. Further, the packet does not indicate the nature of the enclosed payload like a protocol ID or whatsoever. Q3P requires the parties listening on the channels to understand and interpret the given data and hands it out as-is.

8.4 Routing

With the help of Q3P, point-to-point connections are established which are capable of sending and receiving messages in an information theoretical secure manner. Q3P is now capable to encapsulate protocols; one can do this with the well-known IP to achieve routing. But for dynamic routing each hop or node has to maintain its routing tables and in order to establish some sort of QoS we have to examine known interior gateway protocols like OSPF and adopt these to QKD needs.

When using OSPF (see [8]) messages about the current state of the network are "announced" by so called LSA[2] packets. These announcements are flood periodically in the network to inform neighbor nodes of the current "best" routes. Each node states its current costs to other nodes/networks it knows. Based on this information the participating nodes can recalculate their routing tables to estimate the shortest paths to known nodes/networks. "Shortest" actually means "cheapest" in terms of cost, whereas the cost function should be flexible, taking key store capacities and key generation rate into account. If no further action is undertaken each and every LSA is authenticated and encrypted, which seems too much:

1. It is unclear to which extend LSA packets have to be encrypted. What can an attacker do with the information about the link states of each QKD node?
2. As key consumption regarding encryption equals message size a constant key consumption rate is ongoing without the network processing user data.

These and other issues are addressed by the QKD-NL – Quantum Key Distribution - Network Layer – protocol (see [2]).

As the messages are passed via Q3P we can choose to only authenticate the messages and not to encrypt them. This already leverages (2) although the background key consumption of the network is not zero, since every LSA packet receives at least an authentication tag. This tag costs for key consumption equals tag size and occurs on every message sent.

We can further tweak background key consumption by shorten or lengthen the announcement timespan. By enlarging the timespan we can spare key material and lower network background key burning. However, the bigger the timespan the slower network changes are propagated through the network and this may result

[2] *L*ink *S*tate *A*nnouncement

in inefficient network usage, since things may change quicker than communicated or lots of failed connection attempt since nodes may not be present any longer as suggested on the ingress side.

As opposed to standard OSPF the new QKD-NL protocol considers more than one path to a destination node. This is done in order to achieve some kind of load balancing. In OSPF the forwarding decision is based solely on the shortest path and thus every packet addressing a certain network will pass through one single port. In QKD-NL, however, a local load balancing is introduced. Along with the standard routing table an *Extended Routing Table* is present, which lists all costs from one node to each target nodes in increasing magnitude of cost. Next a *Load State Database* keeps track of the approximated load of each outgoing link. Whenever a packet is to be transmitted the Load State Database is looked up and if the local load of the outgoing link is below a threshold the packet is transmitted on exactly this link. If the current load on the link exceeds the limit the next best link is looked up and so forth.

8.5 Transport

A QKD network enables two parties, Alice and Bob, to share the same key on an end-to-end connection across several intermediate hosts. After operating jointly on a common view of the QKD devices and the key sources via Q3P and having an optimized routing scheme in the intermediate host given by QKD-NL such end-to-end key distribution systems can be designed.

Based on the well-known TCP the QKD-TL – *Quantum Key Distribution Transport - Transport Layer* – protocol addresses

- Reliability
- Congestion control
- Connection management
- Multiplexing network services

QKD-TL packets are enclosed in QKD-NL packets, which in turn are enclosed in Q3P packets. As QKD-TL orients itself tightly to TCP (as QKD-NL does with IP) the main differences to TCP are described here. The principle of QKD-TL is to pick a random number, preferentially from a true random number source, and send it hop-by-hop across the network. On the receiver side this random number is treated as key. Superfacial this can be achieved already with TCP, but differences do exist.

First, the messages which are transmitted over this network do not have arbitrary content. These are quantum generated keys as depicted in Sect. 8.1 and – as such – random numbers. For every hop an additional key plus the key material for an authentication tag is consumed. Now, whenever some node along the transmission path has to drop such message due to congestion pure TCP results in a timeout on the sender side, halving the windows size and resending the packets starting from the last good acknowledgment.

In QKD-TL things are different. Since we do not know where the packet has been dropped we might sacrifice a lot of key material for classical TCP resend. Even worse: the key might have been delivered but the corresponding ACK packet got lost. Now resending the whole key again for no additional information benefit besides the sender knowing the key has already reached the receiver is not optimal. Based on the fact, that within QKD-TL a message content is actual a series of pure random bits without any further informational impact or application intention besides using this data as a cryptographic key, there is no resend mechanism. This has a deep impact in subprotocols to follow.

Congestion deals with the problem on how to lower network traffic in such a way that every participating party has a fair amount of bandwidth. The resources in question are the QKD keys needed for transport in each intermediate host. The technique to calculate the reserves is to watch the outgoing key stores of Q3P plus the approximated key generation rate on a single Q3P link. If these values drop below a certain threshold a key shortage is likely to happen. Now QKD-TL introduces a single new Bit in the flags section of the TCP header field: *Congested*. This Bit is set, whenever a node sees its key material on an outgoing line below its label. The destination node suppresses the sending of an ACK packet back whenever this Congested Bit is set in packet it reads. This results in a timeout on the sender side and thus halving the window size which in turn lowers network bandwidth and key consumption.

Missing a resend facility within QKD-TL raises new questions on how to hand out key material. For this, QKD-TL cannot be seen as a stream of key material but has been rather designed utilizing a client–server approach. Figure 8.8 pictures the basic setup. It shows a typical setup: two areas are drawn which share access to a QKD network on the bottom and access to a classical public network on top. On the left side resides a client which wants to share a key with a server on the right side. This scenario is true for companies having several branches connecting to some central unit running various application servers.

Itemizing the steps in a simplified manner:

1. The application server registers itself within the QKD network and states it readiness to accept key requests by clients through the QKD network. It does this registration on its next QKD network node running a QKD-TL connection manager.
2. A client contacts its nearest QKD network node and requests the residing QKD TL connection manager there to establish a connection with the server.
3. Throughout the QKD network a connection is established, session IDs exchanged. Along these sessions, the client side QKD-TL manager starts generating keys along with a key ID, sends them across the QKD network to the server's QKD-TL manager. The latter hands the keys out to the server and acknowledges the key reception. On receiving the acknowledgment the client's QKD-TL manager hands out the key to the client.
4. Based on the session ID and on the key ID the client and the server negotiate keys for concrete application usage.

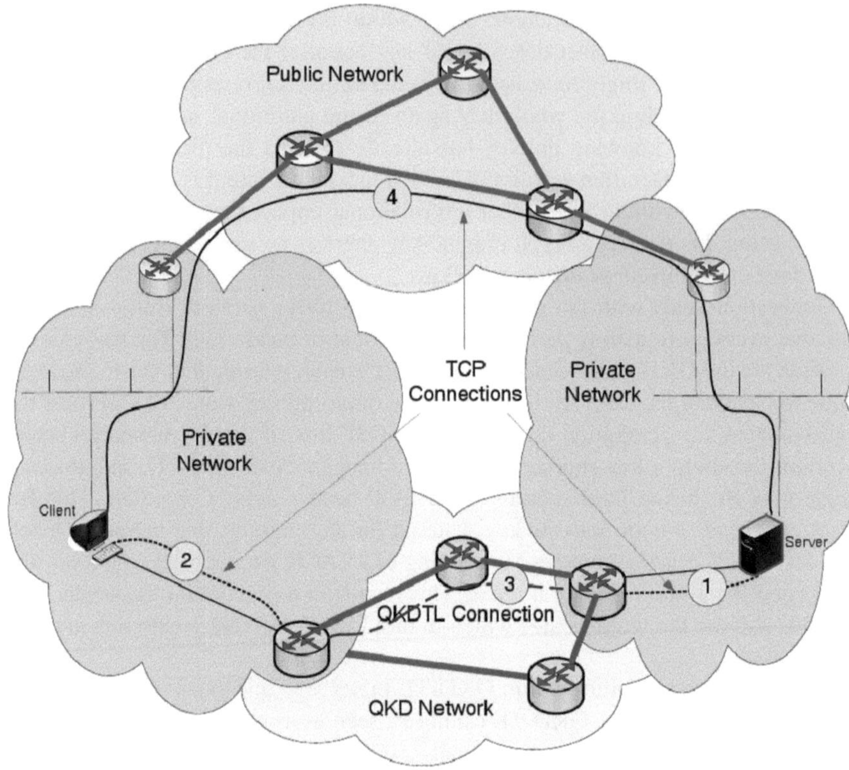

Fig. 8.8 QKD-TL and client-server (source: [2])

As QKD-TL does not have a resend mechanism any key transmitted from the client's side which lacks a corresponding acknowledgment may be lost. QKD-TL itself does not provide a key handshaking protocol yet but whenever one is going to be designed this idiosyncrasy prevails: due to the fact that not the key packet but an ACK packet may have been lost in the QKD network, the server may have keys the client lacks at his disposal. Therefore, any key negotiation must ground on client key material for the some of the server's keys have to be dismissed.

References

1. Bennett, C.H., Brassard, G.: Quantum cryptography: Public key distribution and coin tossing. In: Proceedings of the IEEE International Conference on Computers, Systems, and Signal Processing, pp. 175–179 (1984)
2. Dianati, M., Allaume, R., Gagnaire, M., Shen, X.S.: Architecture and protocols of the future European quantum key distribution network. In: Security and Communication Networks. John Wiley & Sons, New York (2009)
3. Gross, G., Kaycee, M., Li, A., Malis, A., Stephens, J.: RFC 2364: PPP over AAL5 (1998). URL ftp://ftp.internic.net/rfc/rfc2364.txt, ftp://ftp.math.utah.edu/pub/rfc/rfc2364.txt

4. Hamzeh, K., et al.: RFC 2637: Point-to-point tunneling protocol (pptp) (1999). URL http://www.ietf.org/rfc/rfc2637.txt
5. Kent, S., Atkinson, R.: RFC 2401: Security architecture for the Internet Protocol (1998). URL ftp://ftp.internic.net/rfc/rfc1825.txt, ftp://ftp.internic.net/rfc/rfc2401.txt, ftp://ftp.math.utah.edu/pub/rfc/rfc1825.txt, ftp://ftp.math.utah.edu/pub/rfc/rfc2401.txt
6. Mamakos, L., et al.: RFC 2516: A method for transmitting ppp over ethernet (pppoe) (1999). URL http://www.ietf.org/rfc/rfc2516.txt
7. McGregor, G.: RFC 1332: The PPP Internet Protocol Control Protocol (IPCP) (1992). URL ftp://ftp.internic.net/rfc/rfc1332.txt, ftp://ftp.math.utah.edu/pub/rfc/rfc1332.txt
8. Moy, J.: RFC 2328: Ospf version 2 (1998). URL ftp://ftp.internic.net/rfc/rfc2364.txt, ftp://ftp.math.utah.edu/pub/rfc/rfc2364.txt
9. Poppe, A., Peev, M., Maurhart, O.: Outline of the secoqc quantum-key-distribution network in vienna. In: International Journal of Quantum Information IJQI, Vol. 6. World Scientific Publishing (2008). URL http://arxiv.org/abs/0804.0122
10. Postel, J.: RFC 793: Transmission control protocol (1981). URL ftp://ftp.internic.net/rfc/rfc793.txt, ftp://ftp.math.utah.edu/pub/rfc/rfc793.txt
11. Romkey, J.L.: RFC 1055: Nonstandard for transmission of IP datagrams over serial lines: SLIP (1988). URL ftp://ftp.internic.net/rfc/rfc1055.txt, ftp://ftp.math.utah.edu/pub/rfc/rfc1055.txt
12. Simpson, W.: RFC 1661: The point-to-point protocol (PPP) (1994). URL ftp://ftp.internic.net/rfc/rfc1548.txt, ftp://ftp.internic.net/rfc/rfc1661.txt, ftp://ftp.internic.net/rfc/rfc2153.txt, ftp://ftp.internic.net/rfc/std51.txt, ftp://ftp.math.utah.edu/pub/rfc/rfc1548.txt, ftp://ftp.math.utah.edu/pub/rfc/rfc1661.txt, ftp://ftp.math.utah.edu/pub/rfc/rfc2153.txt, ftp://ftp.math.utah.edu/pub/rfc/std51.txt
13. Zimmermann, H.: Osi reference model–the iso model of architecture for open system interconnection. IEEE Transactions Communications Vol. COM-28 (1980). URL http://www.comsoc.org/livepubs/50_journals/pdf/RightsManageme% nt_eid=136833.pdf

Chapter 9
Quantum-Cryptographic Networks from a Prototype to the Citizen

P. Schartner and C. Kollmitzer

9.1 The SECOQC Project

Secure communication is an essential need for companies, public institutions, and in particular the individual citizen. Currently used encryption systems are vulnerable due to the increasing power of computer technology, the emergence of new code-breaking algorithms, and the imperfections of public key infrastructures. Methods considered as acceptably secure today will have a significant risk of becoming weak tomorrow. On the other hand, with quantum cryptography, a technology has been developed within the last decade that is provably secure against arbitrary computing power, and even against quantum computer attacks. When becoming operational, quantum cryptography will raise communication security to an essentially higher level.

The vision of SECOQC (Development of a Global Network for Secure Communication based on Quantum Cryptography, [15, 1, 12, 2, 3]) was to provide European citizens, companies, and institutions with a tool that allows facing the threats of future interception technologies, thus creating significant advantages for European economy. With SECOQC the basis was laid for a long-range high-security communication network that combines the entirely novel technology of quantum key distribution with components of classical computer science and cryptography.

Within the project the following goals were defined:

1. Realization of a fully functional, real-time, ready-to-market quantum key distribution (QKD) point-to-point communication technology (see Chap. 3).

P. Schartner (✉)
System Security Research Group, Institute of Applied Informatics, Universitaet Klagenfurt, Universitaetsstrasse 65-67, 9020 Klagenfurt, Austria, peter.schartner@uniklu.ac.at
http://www.syssec.at

C. Kollmitzer
Safety & Security Department, Quantum Technologies, AIT Austrian Institute of Technology GmbH, Lakeside B01A 9020 Klagenfurt, Austria, christian.kollmitzer@ait.ac.at; http://www.ait.ac.at

Schartner, P., Kollmitzer, C.: *Quantum-Cryptographic Networks from a Prototype to the Citizen*. Lect. Notes Phys. **797**, 173–184 (2010)
DOI 10.1007/978-3-642-04831-9_9 © Springer-Verlag Berlin Heidelberg 2010

2. Development of an abstract level architecture, allowing high-security long-range communication by integrating the QKD technology and a set of cryptographic protocols.
3. Design of a real-life, user-oriented network for practical implementation of QKD-based long-range secure communication.

To achieve this goal, all experiences and resources available within the European Research Area were integrated and combined with the expertise of developers and companies within the fields of network integration, cryptography, electronics, security, and software development [16].

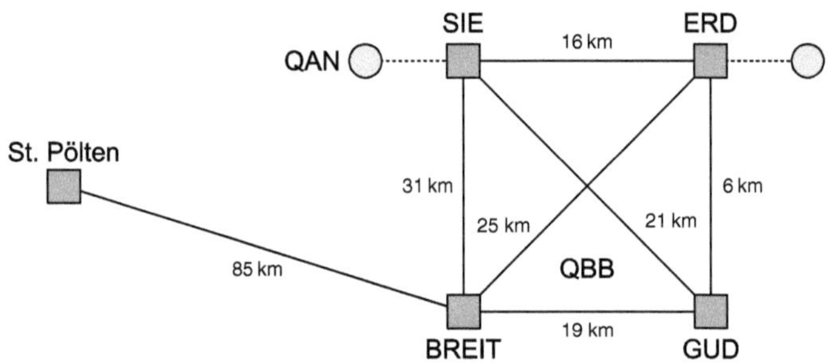

Fig. 9.1 SECOQC Network Vienna

9.1.1 SECOQC Network Vienna 2008

Figure 9.1 shows the concept of the deployed universal building block for QKD networking from [14]. The names refer to the stations of the ring network.

Figure 9.2 shows a map of the fiber ring network in Vienna including the stations SIE (Siemensstraße), ERD (Erdberg), GUD (Gudrunstraße), and BREIT 32 (Breitenfurterstraße) from [14]. Another station, St. Pölten, was located far outside of Vienna.

The SECOQC network was built by using the telecom fiber network from Siemens Austria. As shown in Fig. 9.2 it was built to connect the office buildings of Siemens Austria located in Vienna. The fiber ring connected four nodes (SIE, ERD, GUD, and BREIT) and had a circumference of approximately 85 km. For the SECOQC network, there was also another fiber of about 63 km which connected the city of St. Pölten to the fiber ring. Each connection between two nodes was realized by a QKD device pair which was connected by its own dark fiber. Another fiber loop was used for the classical communication.

The used QKD devices (for details see Sect. 6.2) had to fulfill the following requirement: a stable secure key rate after authentication of at least 1 kbit/s over 25 km of fibers. The systems of the following organizations have met these criteria and took part in the SECOQC network:

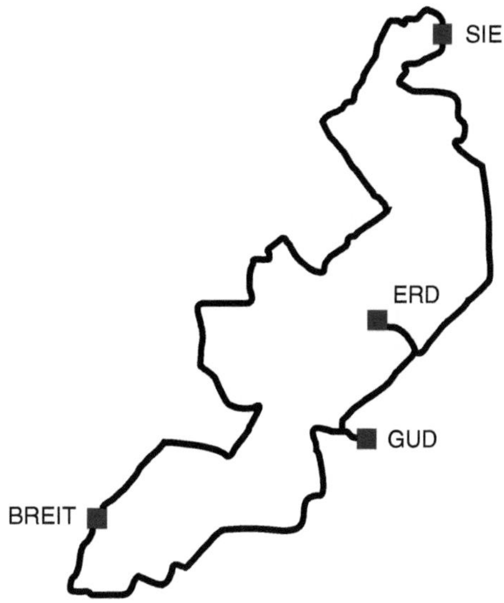

Fig. 9.2 City map of Vienna including the fiber ring network

- IdQuantique (Switzerland) [8],
- GAP Geneva, IdQuantique, ARC (Switzerland, Austria),
- Toshiba (UK),
- University of Vienna, ARC, KTH (Austria, Sweden), and
- CNRS, Thales, ULB (France, Belgium).

9.1.2 Design of a QKD Network

The first practical quantum key distribution took place in 2004 when the headquarters of The Bank Austria Creditanstalt and the Vienna City Hall established a secure communication by using QKD devices. The beeline distance between the communication partners was about 650 m; the installed optical fibers had a length of about 1.45 km [13].

As former QKD schemes, which used to be one-to-one connections, evolve into QKD networks, new network architectures had to be developed. With the SECOQC project, a schema was presented, which consists of two main elements (see Fig. 9.3):

1. A quantum backbone network (QBB), which correlates to a classical backbone network architecture, enhanced for QKD needs like special key stores (see Sect. 8.3.4)
2. A quantum access node (QAN) which allows the user to get the keys, shared with other users within the network.

For more information on global scale QKD-based networks see Chap. 10.

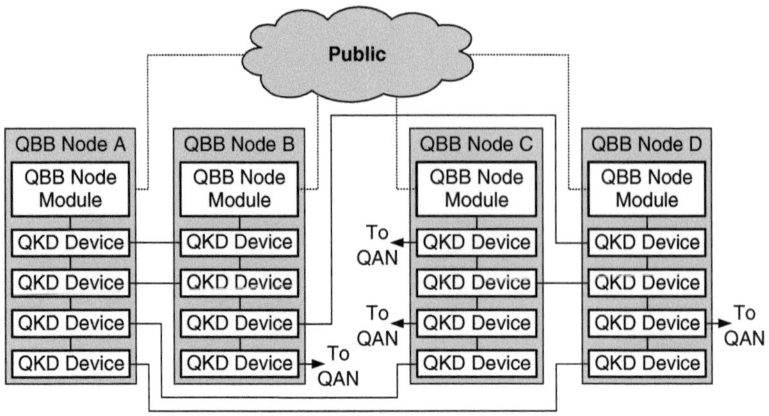

Fig. 9.3 Network architecture used within the SECOQC project

9.2 How to Bring QKD into the "Real" Life

The QBB network at first hand provides one-time pad (OTP) keys for end-to-end security between QBB nodes. In the remainder of this chapter, keys "generated" and used within the QBB network will be called "QKeys." A QKey may be a block of bits, or a stream of bits. In this section, we will discuss some problems which arise, when we want to provide QKeys for everybody. In more detail, we will discuss

- Secure transmission
- Secure storage
- Efficient usage of QKeys

The ideal scenario is that everybody's PC is directly connected to a QAN (Quantum Access Node) by a fiber-optic channel, so that we can protect the last mile by the same mechanisms we employ in the QBB network. We do not think that this scenario will be available in the near future. So, in order to use QKeys everywhere and at any time, we need to design some other techniques. As one possible solution for the problem of secure transport, storage, and usage, we propose small mobile devices like PDAs or SmartPhones with integrated smart cards or special micro SD memory cards (e.g., certgate smart card [5]). These mobile devices may either directly use the QKeys within their applications or they may be used to securely transmit the key to the users' PC.

9.2.1 Secure Transmission to the Mobile Device

The QBB network provides a securely transmitted key-stream between two QANs. In order to use these key-streams within applications, they have to be forwarded to end users, more precisely to the end users' devices. QBB provides unconditional security of the key-streams, so the key-streams have to be protected on the last two links (one at the sender's and one at the receiver's end). Ideally, this protection should provide unconditional security, and if we can't afford unconditional security, we should at least employ a state-of-the-art encryption schemes (e.g., symmetric encryption by means of AES) and some high-entropy key. In this section we will discuss several methods of protecting the last link. Analogous to high-speed Internet we will discuss methods for bridging the last mile (which isn't really a mile here) with existing technologies.

Choosing the appropriate method of data transmission isn't an easy task. Today's mobile devices most commonly provide several communication technologies or input interfaces. Table 9.1 summarizes the advantages and disadvantages of these communication technologies.

The best method for wireless key transmission is obviously freespace QKD. In addition to this method, we will discuss two alternatives. One which uses Near Field Communication (NFC [6]) and another which uses the camera of the mobile device to transmit the session key or the QKey.

Table 9.1 Input interfaces – Pros and Cons

Interface	PROs	CONs
Wireless (QKD)	• unconditional secure	• low bit rate • cumbersomely to use • quite clumsy and expensive
Wireless (RF)	• easy to use • high speed	• sniffing hard to detect • hard to shield
Wireless (IR)	• easy to shield	• difficult alignment • medium bit rate
Wired	• easy to use • high speed	• location of interface not standardized • hidden sniffers hard to detect • vandalism
Audio	• widely supported • easy to use	• hard to shield • low bit rate
Video	• easy usage and alignment • unidirectional communication • easy to shield	• medium bit rate • exotic way of transmission
Text	• widely supported • unidirectional communication	• cumbersomely to use • very low bit rate

9.2.1.1 Quantum Key Distribution – QKD

Ideally, the key material will be transmitted encrypted by use of an unconditionally secure encryption algorithm. Up to now, the one-time pad is the only such algorithm. To employ the one-time pad, we need a key which is as long as the message. Additionally, this key must be random and must not be used twice. So it is obvious, that QKD is the first choice in order to provide unconditional security over all hops, especially the last one. Unfortunately at the time of writing, there are no cheap and small (i.e., size of a standard mobile phone) QKD devices. Additionally, mobile devices should ideally be linked by wireless transmission, in case of QKD this means freespace transmission.

At the time, freespace QKD provides a bit rate of approximately 10–15 kbit/s at a distance of about 80 m [17]. Note that this bit rate can only be achieved under ideal circumstances. A less exact alignment of sender and receiver or atmospheric influences can easily reduce the bit rate.

Using QKD (over fibre optics or freespace) to protect the last mile is desirable, but (at the time) not that practical. So if we do not want to wait, we need other techniques. The second-best choice after OTP encryption with a QKD key is OTP encryption with a high-entropy key exchanged by classical mechanisms. If we need high throughput, we might use a state-of-the-art encryption algorithms and a key of appropriate length. Nevertheless, in all these cases, we need to exchange a session key.

9.2.1.2 Near Field Communication – NFC

In [7], Haselsteiner and Breitfuss proposed an interesting key agreement scheme for Near Field Communication (NFC [6]), which does not involve any security mechanisms at all. The protocol is run between two users A and B, the adversary will be denoted by E. In principle, A and B repeat the following steps until a sufficient amount of key bits has been generated.

1. Both, A and B, generate a random bit (b_A and b_B, respectively) which is sent simultaneously to the other party. In parallel both listen to the communication channel. Since they both send one bit, there are four possible combinations, which are shown in Fig. 9.4. Here, a 1 is represented by sending in the first half of the time slot, whereas a 0 is represented by sending in the second half of the time slot.

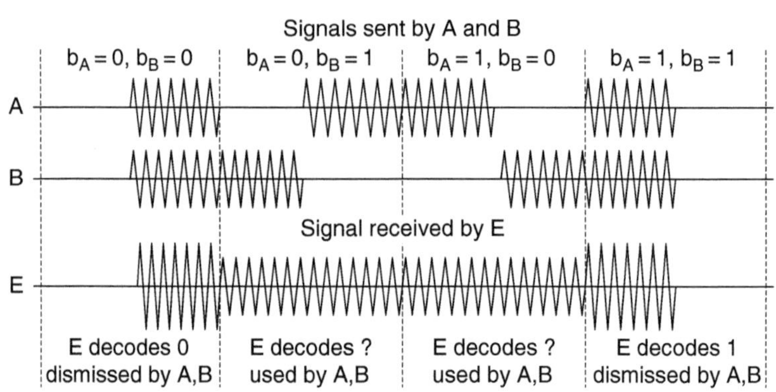

Fig. 9.4 NFC key agreement

2a If $b_A = b_B$, *A* and *B* can detect this easily, and so can the attacker *E*. Hence, *A* and *B* have to dismiss their bits and continue with step 1.

2b If *A* and *B* sent different bits, they can detect this easily, too. But now, the attacker only knows that *A* and *B* sent different bits. He does not know if $b_A = 0$ and $b_B = 1$ or $b_A = 1$ and $b_B = 0$. So *A* and *B* can use either the bit sent by *A* or the bit sent by *B* to generate a session key. They only need a strategy, whose bit will be used. After adding the selected bit to the session key, *A* and *B* continue with step 1.

Note that this describes only the operating principle. In order to provide a high level of security, the sending units of *A* and *B* have to be "absolutely" synchronous. Additionally, the amplitudes of the signals generated by *A* and *B* have to be indistinguishable.

Since the NFC key agreement uses radio transmission, we have to provide some electromagnetically shielded environment. This raises the question "Why not transmit the key in clear?" or, for the sake of security, encrypted by use of a rather short key and high-speed symmetric encryption scheme?

9.2.1.3 Optical Transmission

When we think of wireless transmission of the key-stream, we should not restrict ourselves to radio frequency (i.e., WLAN, Bluetooth, of NFC). Since almost all current phones and most PDAs are equipped with a camera, we might use optical transmission of the QKeys instead of radio transmission. Again, this is not the ideal (i.e., high-speed, unconditional-secure) mechanism. But it is cheap, easy to implement, and easy to use! Additionally, optical transmission is much easier to shield than radio frequency transmission (light doesn't propagate around corners that easily).

So how to apply optical transmission between the key terminal and the mobile device? One method to encode the data is so-called QR codes (quick response codes) [9, 10, 11]

Figure 9.5 shows some QR codes which have been generated by use of the free tool of KAYWA AG, Switzerland (http://qrcode.kaywa.com). The left one encodes

Fig. 9.5 QR codes

the URL “www.secoqc.net,” the middle one encodes the text “SECOQC – Development of a Global Network for Secure Communication based on Quantum Cryptography,” and the right one encodes the first 236 characters of the first paragraph of the Wikipedia page on QR codes (“http://en.wikipedia.org/wiki/QR_Code”). These sample QR codes correspond to about 77, 528, and 1.298 bits, respectively (~5.5 bits per alphanumerical character).

According to Denso Wave Incorporated, QR codes – which use Reed-Solomon for error detection and error correction – come with the following maximum data capacity:

- 1.817 Kanji/Kana characters, or
- 4.296 alphanumeric characters, or
- 7.089 numeric only characters, or
- 2.953 bytes (8 bits each, so 23.624 bits in total).

Most PDA and Smartphone cameras come with a minimal focus distance of about 30–50 cm. This means that the LCD screen displaying the QR codes has to be in similar distance to the camera (see the left side of Fig. 9.6 for a sketch of such a system). In order to reduce the size of the key terminal, we could use the camera to detect the presence or absence of light only (binary coding). In this case, the minimal focusing distance doesn’t matter and we can shrink the size of the key terminal to a minimum.

In order to overcome the reduced bandwidth of binary transmission, we could split the camera image into several subsections (each detecting the presence/absence of light) or we could use several colors like indicated in the right half of Fig. 9.6. In both cases the bandwidth will be increased by 2^n, where n is the number of sections or colors, respectively. Additionally, the images are now much simpler than QR codes and hence can be decoded much faster.

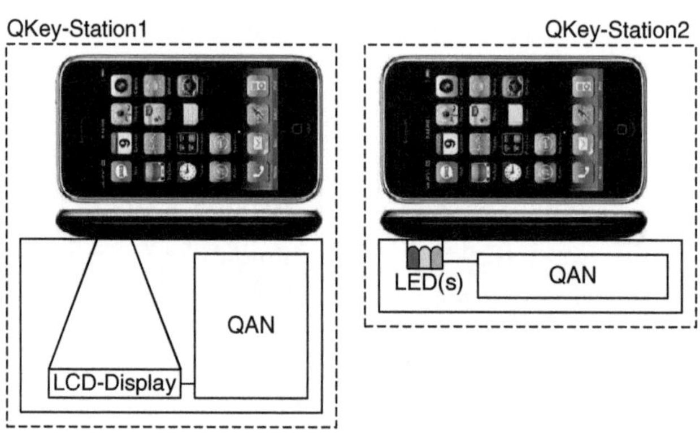

Fig. 9.6 iPhone 3G at a key terminal

9.2.2 Secure Storage

In our scenario, there is no online connection between the QAN and the mobile device. So it is obvious that we have to securely store an appropriate amount of key material within the mobile device over a longer period of time. So let us briefly analyze some types of mobile devices with respect to secure storage:

- *Laptops and Netbooks:* Indisputably, these devices provide a huge storage capacity, but they are quite clumsy and very hard to secure.
- *PDAs and SmartPhones:* These devices are rather handy and come with sufficient memory, but are still quite hard to secure.
- *Smart Cards:* Smart Cards come with two flavors: smart cards (also known as ICCs – integrated circuit cards) and special secure memory cards.
  - *Smart Cards or ICCs:* By now, there are four standardized formats: ID-0 (credit card size), ID-00, ID-000 (plugin or SIM size), and Mini-UICC (about half plugin-size). These cards can provide up to 1 GB of non-volatile memory and most commonly come with a cryptographic co-processor (e.g., 2048 RSA in 10 ms, AES in 10 μs). Besides these positive features, smart cards come with a major disadvantage: that we need some type of special terminal in order to connect them to a PC, Laptop, PDA, or SmartPhone.
  - *Secure memory cards:* Within memory cards, a special type of SD memory crads (short SDCards [4]) is quite promising: the certgate SDCard [5]. This SDCard (1 GB) is provided with a smart card microprocessor. Since this processor can be used over the standard SDCard interface, no special reader is needed.

Up to now, Personal Digital Assistants (PDAs) or SmartPhones which are equipped with a smart card or some other security token (e.g., a certgate SDCard) will fulfill our security requirements. It is obvious that this type of equipment is not ideal (e.g., it is not unconditional secure), but it is available, quite cheap, widely adopted and accepted by the customers.

9.2.3 Efficient Key Usage

In the SECOQC design, a standard QBB node has three types of key buffers (called KeyStores):

- *In-KeyStore:* Keys within this KeyStore are reserved for incoming messages of the QBB network.
- *Out-KeyStore:* Keys within this KeyStore are reserved for outgoing messages of the QBB network.
- *Application KeyStore:* Keys within this KeyStore are handed over to external applications.

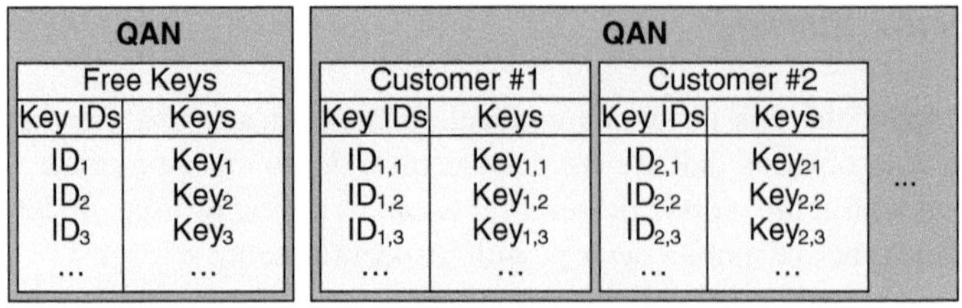

(a) Free Keys (b) Customer/Application specific Keys

Fig. 9.7 Quantum access node keystore

In order to guarantee a minimum number of QKeys for special applications, we would like to propose an additional component which divides the keys of the Application KeyStore into two classes: so-called free keys and designated keys (see Fig. 9.7. The IN- and OUT-buffer of the QBB node will remain reserved for internal use (i.e., key forwarding).

- *Free Keys:* Free keys may be used for any purpose (i.e., encryption or authentication of any data type) and any receiver. After retrieval of the communication key at the sender's end, the system does not know the receivers address. So there are at least two possible strategies:

  1. The sender already knows the receiver (respectively, his address). Now, the receiver can be informed by the system and the receiver can immediately retrieve the communication key.
  2. The sender does not want to send data right now, he simply wants to get some keys for his local key storage. Now there can't be done anything in advance at the receiver's end. The receiver has to wait for the encrypted/authenticated message in order to identify the used key. Now he can retrieve the key and decrypt/check the message.

- *Designated Keys:* If a sender retrieves a key reserved for a special receiver (or purpose), the designated receiver can automatically be informed as soon as the key has been retrieved by the sender. Hence, the receiver is able to fetch his key at the moment, the sender requests it from the KeyStore.

9.3 Resumee

So far, the field of QKD has been analyzed from several different points of view. Especially, networks have been in the center of research interest. With the SECOQC demonstration in October 2008 it has been shown that networks based on QKD are realizable, which changed the focus on the security research. On one hand there will be growing interest on QKD-based applications and, on the other hand, we expect an increasing research activity on the enhancement of already existing communication infrastructures regarding a secured key exchange. In our opinion, transmission, storage, and usage of keys generated by QKD becomes the new emphasis of research work.

In order to achieve an appropriate market throughput, QKD systems will have to offer solutions for different security levels. Thereby, different areas must be covered, e.g., QR codes offer the possibility of a fast, widely spread, and easy to use method for applications of a lower security level.

From our point of view, QKD is a massive future topic. The research focus will spread in the next years and not only the physical fundamentals will be examined but also the aspects of applications will be more and more of interest. We think in particular that scenarios with simple, easy to access interfaces, designed for a large number of users such as telephone boxes or ATM-like systems, will be of strong interest.

References

1. Alléaume, R., Bouda, J., Branciard, C., Debuisschert, T., Dianati, M., Gisin, N., Godfrey, M., Grangier, P., Länger, T., Leverrier, A., Lütkenhaus, N., Painchault, P., Peev, M., Poppe, A., Pornin, T., Rarity, J., Renner, R., Ribordy, G., Riguidel, M., Salvail, L., Shields, A., Weinfurter, H., Zeilinger, A.: SECOQC white paper on quantum key distribution and cryptography (2007). http://www.citebase.org/abstract?id=oai:arXiv.org:quant-ph/07%01168
2. Alléaume (Editing author), R., Bouda, J., Branciard, C., Debuisschert, T., Dianati, M., Gisin, N., Godfrey, M., Grangier, P., Länger, T., Leverrier, A., Lütkenhaus, N., Painchault, P., Peev, M., Poppe, A., Pornin, T., Rarity, J., Renner, R., Ribordy, G., Riguidel, M., Salvail, L., Shields, A., Weinfurter, H., Zeilinger, A.: Quantum key distribution and cryptography. SECOQC White Paper (2007)
3. Assche, G., Cardinal, J., Cerf, N.: IEEE Trans. Inf. Theory **50**, 394 (2004)
4. Association, S.: (2009). http://www.sdcard.org
5. Certgate: Certgate Secure Digital Card (2009). www.certgate.de
6. Forum, N.: (2009). http://www.nfc-forum.org/home
7. Haselsteiner, E., Breitfuss, K.: Security in near field communication (NFC). Hand-out of Workshop on RFID Security RFIDSec06 (2006). http://events.iaik.tugraz.at/RFIDSec06/Program/papers/002%20-%20Security%20in%20NFC.pdf
8. IdQuantique: (2009). http://www.idquantique.com
9. Inc., D.W.: QR-Code Standardization (2009). http://www.denso-wave.com/qrcode/qrstandard-e.html
10. ISO: (2006). ISO/IEC 18004:2006 – Information technology – Automatic identification and data capture techniques – QR Code 2005 bar code symbology specification
11. ISO: (2009). ISO/IEC 18004:2006/Cor 1:2009
12. Monyk (Coordinator), C.: Development of a global network for secure communication based on quantum cryptography. EC/IST Integrated Project SECOQC **Contract No. 506813** (2004-2008)
13. Poppe, A., Fedrizzi, A., Ursin, R., Böhm, H.R., Lorünser, T., Maurhart, O., Peev, M., Suda, M., Kurtsiefer, C., Weinfurter, H., Jennewein, T., Zeilinger, A.: Practical quantum key distribution with polarization entangled photons. Opt. Express 3865 **12** (2008)
14. Poppe, A., Peev, M., Maurhart, O.: Outline Of The SECOQC quantum-key-distribution network in Vienna. Int. J. Quantum Inf. **6**(2) (2008)
15. SECOQC – Development of a Global Network for Secure Communication based on Quantum Cryptography: (2009). http://www.secoqc.net

16. Sheet, I.P.F.: (2009). http://cordis.europa.eu/fetch?CALLER=PROJ_IST&ACTION=D&RCN=71407
17. Ursin, R., Tiefenbacher, F., Schmitt-Manderbach, T., Weier, H., Scheidl, T., Lindenthal, M., Blauensteiner, B., Jennewein, T., Perdigues, J., Trojek, P., Ömer, B., Fürst, M., Meyenburg, M., Rarity, J., Sodnik, Z., Barbieri, C., Weinfurter, H., Zeilinger, A.: Free-Space distribution of entanglement and single photons over 144 km (2006)

Chapter 10
The Ring of Trust Model

C. Kollmitzer and C. Moesslacher

10.1 Introduction

The following chapter deals with a possible application of methods of quantum cryptography to permit secure communication between parties in different architectures by establishing a Ring of Trust. The aim is to solve the problems of key distribution with methods of quantum cryptography without being limited by their transmission range. At the same time, a high level of security is obtained, which is ensured by the use of corresponding cryptographic algorithms. The model is not restricted to certain cryptographic algorithms. Therefore, it is possible to enlarge a preexisting system with new cryptographic algorithms or to replace formerly employed cryptographic algorithms with new ones.

The model presented here limits neither the number of communication parties nor the maximum distance between the different parties. Quantum key distribution (QKD) has now reached a stage of maturity which makes it possible to implement QKD into existing infrastructures, for example, in medical information systems (MIS), and thus enhances their security level significantly. Medical information systems are a critical field because patient-related data and surgery data are both confidential and a subject to legal restraints. Furthermore every kind of medical data has to be highly available.

There also arise new services from this field which can only be implemented if a secure transmission and storage of data can be guaranteed.

C. Kollmitzer (✉)
Safety & Security Department, Quantum Technologies, AIT Austrian Institute of Technology GmbH, Lakeside B01A 9020 Klagenfurt, Austria,
christian.kollmitzer@ait. ac.at; http://www.ait.ac.at

C. Moesslacher
Safety & Security Department, Quantum Technologies, AIT Austrian Institute of Technology GmbH, Lakeside B01A, 9020 Klagenfurt, Austria, christian.moesslacher@gmx.net

Kollmitzer, C., Moesslacher, C.: *The Ring of Trust Model*. Lect. Notes Phys. **797**, 185–210 (2010)
DOI 10.1007/978-3-642-04831-9_10 © Springer-Verlag Berlin Heidelberg 2010

10.2 Model of the Point of Trust Architecture

Two communication parties who want to communicate can be in two different Trust Zones, as shown in Fig. 10.1. In order to communicate they have to rely on a Point of Trust they are assigned to. Each client is connected to its Point of Trust via a quantum channel. All clients of a single Point of Trust form a Trust Zone. Different Points of Trust are connected to each other either directly or via other Points of Trust using either a public channel or a quantum channel. Also the different clients are connected via a public channel, but in order to have a secure communication beyond their Trust Zone they have to establish a session key via their Points of Trust and the corresponding network.

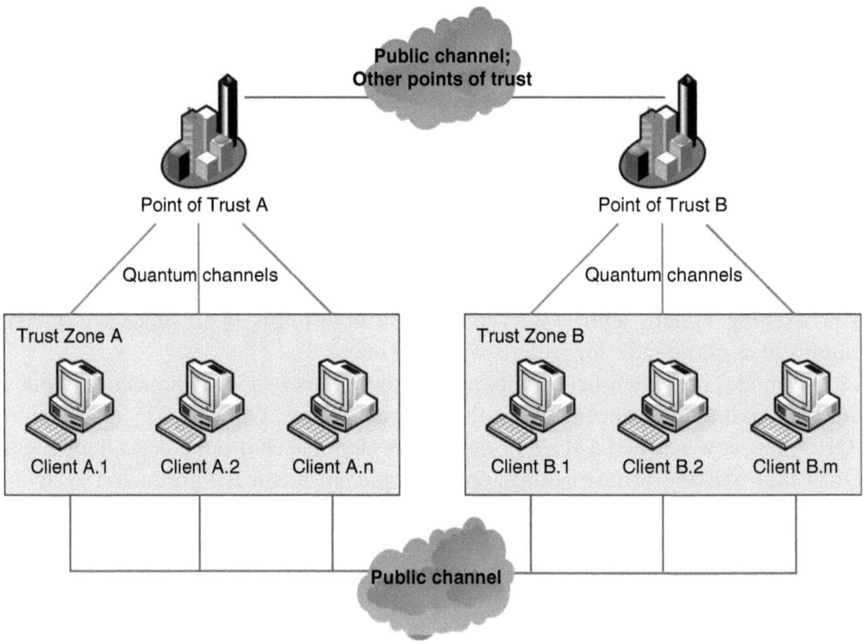

Fig. 10.1 Point of Trust architecture with several Trust Zones

If both communication parties are within one Trust Zone, as shown in Fig. 10.2 (i.e., both are connected to the same Point of Trust) it is easy for them to establish a secure communication via the assigned Point of Trust. But to be able to communicate with both, communication parties inside and outside their own Trust Zone, they can use the Ring of Trust architecture.

10.3 Communication in the Point of Trust Model

The following examples describe the secure communication between two clients. These two clients are either in different Trust Zones, as outlined in 10.3.1.1 and 10.3.2.1, or in the same Trust Zone, as described in 10.3.1.2 and 10.3.2.2. Due to

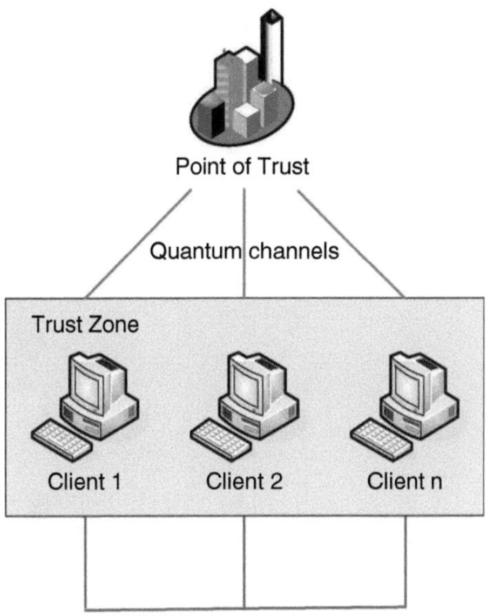

Fig. 10.2 Point of Trust architecture with one Trust Zone

the new architecture presented in this chapter, it is no longer necessary for the two legitimate communication parties to exchange keys over a secure channel. Yet, they can still communicate at a high level of security.

10.3.1 Resource-Oriented Setup of Communication

Contrary to the speed-oriented setup as described in 10.3.2, both steps are triggered by the initiator. Therefore, more steps must be carried out than in the speed-oriented setup.

10.3.1.1 Point of Trust Architecture with Several Trust Zones

In this architecture, communication takes place between clients of different Trust Zones. The participating Points of Trust must be trustworthy, i.e., they must have exchanged a key. Such a network of Points of Trust can be set up in different ways, e.g., in the form of a hierarchical model or a point-to-point model as shown in Figs. 10.3 and 10.4. The form of communication between the respective Points of Trust is not relevant for the function of the Point of Trust model.

Step 1: Initiation of communication

At the beginning of the communication the situation is given by the following conditions. Client A.1 requests client B.2 to communicate and client A.1 and client B.2 are members of different Trust Zones.

The initialization of communication is shown in Fig. 10.5. The communication steps are delineated subsequently.

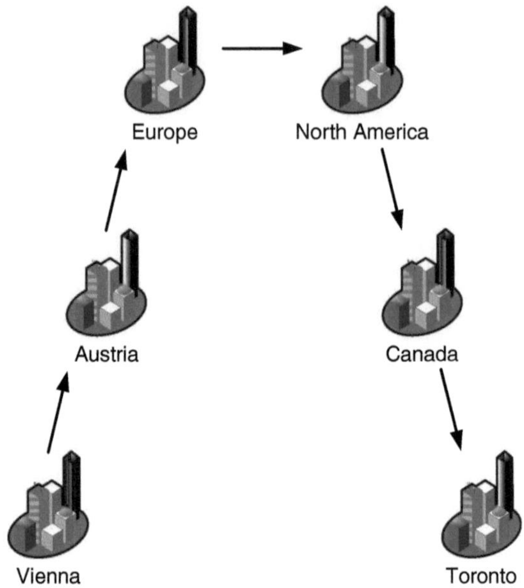

Fig. 10.3 Hierarchical model

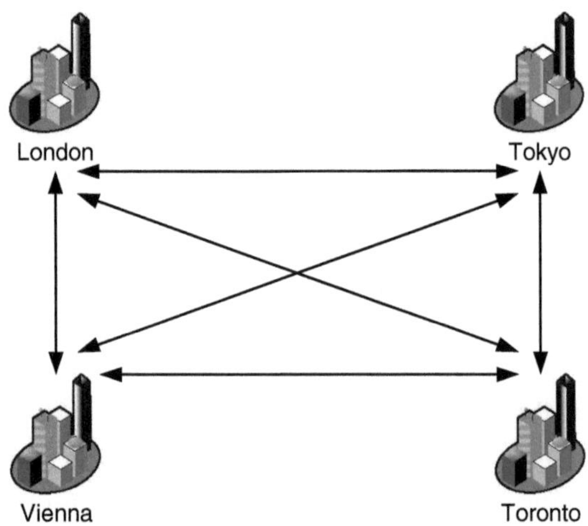

Fig. 10.4 Point-to-Point model

To initialize the communication client A.1 sends a request and transmits its data (authenticity, authorization, accounting information, etc.) to Point of Trust A (*I*) in order to be identified as part of Trust Zone A. Point of Trust A checks the data of client A.1. After positive validation, the request for communication party client B.2 is transmitted to Point of Trust B (*A1*) that is responsible for the target, client B.2. (Points of Trust A and B might be connected directly or via other Points of Trust).

Point of Trust B receives the request for communication between the communication parties client A.1 and client B.2 and requests the identification of client B.2 to communicate (*A2*). Client B.2 receives the request and sends his data (authenticity,

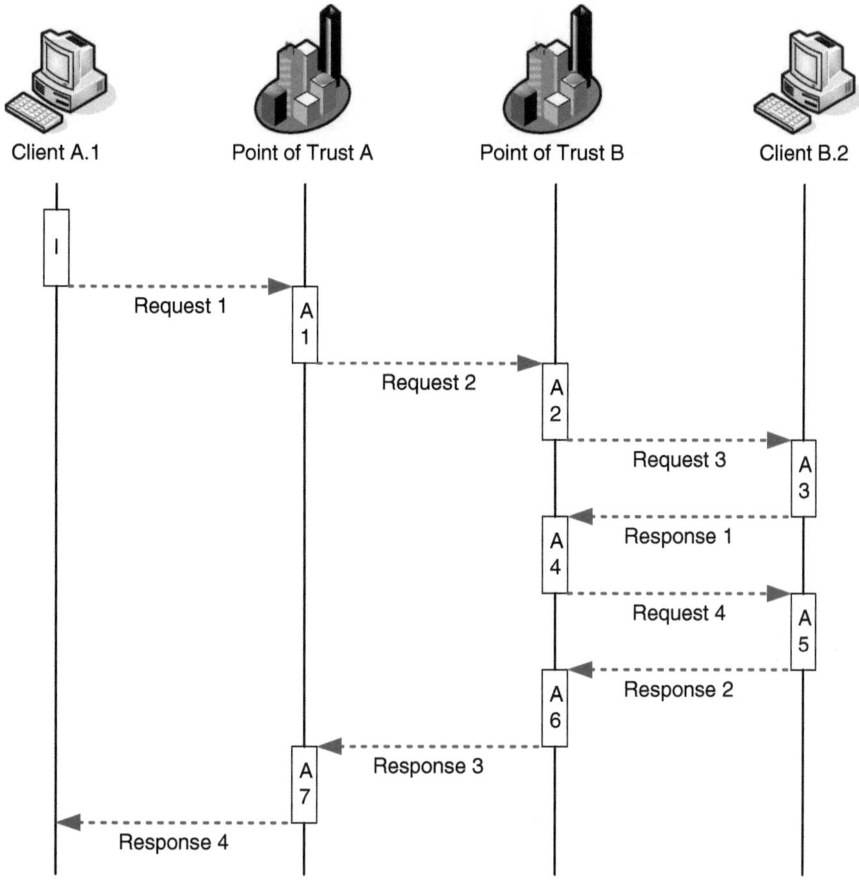

Fig. 10.5 Initiation of communication

authorization, accounting information, etc.) to Point of Trust B as response (*A3*) in order to be identified as part of Trust Zone B. Point of Trust B receives the response and checks the data of client B.2. After positive validation, the information relevant to the communication is sent to client B.2 (*A4*), introducing the requesting communication party, client A.1.

Client B.2 confirms the request for communication from client A.1 and sends his response to Point of Trust B (*A5*). Point of Trust B receives the confirming response and forwards it to Point of Trust A (*A6*). Point of Trust A receives the confirming response and forwards it to the requesting communication party, client A.1(*A7*).

Step 2: Setup of secure communication

The setup of secure communication is shown in Fig. 10.6. The communication steps are delineated subsequently.

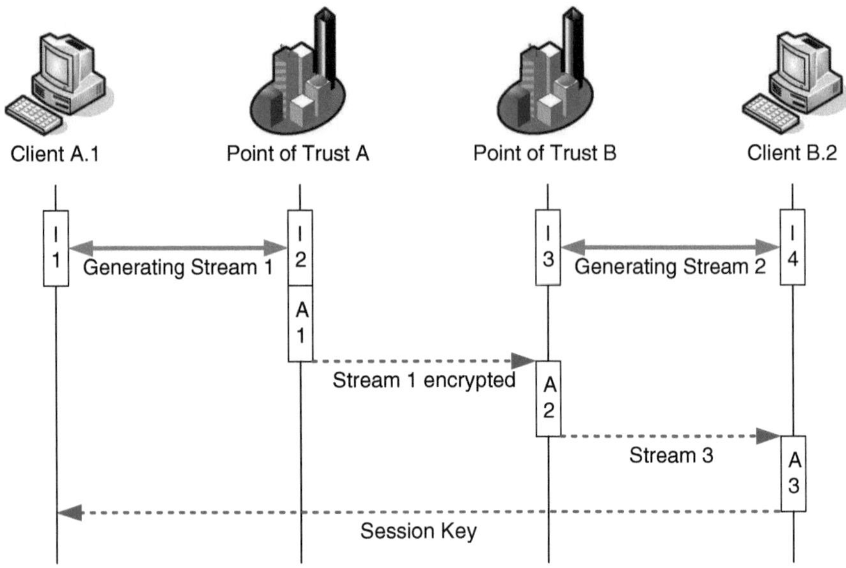

Fig. 10.6 Setup of secure communication

Client A.1 and Point of Trust A are ready to communicate and generate Stream 1, a shared stream generated by using quantum mechanical techniques, over a quantum channel (*I1* and *I2*). Analog, client B.2 and Point of Trust B are ready to communicate and generate Stream 2 (*I3* and *I4*).

Both communication parties have generated streams they share with their respective Points of Trust. Point of Trust A encrypts Stream 1 using a key shared with Point of Trust B generating Stream 1 encrypted and transmits it to Point of Trust B (*A1*). Point of Trust B receives Stream 1 encrypted and uses the key shared with Point of Trust A to decrypt it. The reproduced Stream 1 is then encrypted by Stream 2 generating Stream 3 which is transmitted to client B.2 (*A2*).

Client B.2 receives Stream 3 and uses Stream 2, which is known to him, to reproduce Stream 1. Stream 1 is subsequently established as the session key for the communication parties, client A.1 and client B.2 (*A3*).

10.3.1.2 Point of Trust Architecture with One Trust Zone

Step 1: Initiation of communication

At the beginning of the communication the situation is given by the following conditions. Client A.1 requests client A.2 to communicate and client A.1 and client A.2 are members of the same Trust Zone.

The communication steps for the initialization of communication are delineated subsequently.

To initialize the communication client A.1 sends a request and transmits his data (authenticity, authorization, accounting information, etc.) to Point of Trust A in order to be identified as part of Trust Zone A. Point of Trust A checks the data of client A.1. After positive validation, Point of Trust A requests the identification of client A.2 to communicate.

Client A.2 receives the request and sends his data (authenticity, authorization, accounting information, etc.) to Point of Trust A as response in order to be identified as part of Trust Zone A. Point of Trust A receives the response and checks the data of client A.2. After positive validation, the information relevant to the communication is sent to client A.2, introducing the requesting communication party, client A.1.

Client A.2 confirms the request for communication from client A.1 and sends his response to Point of Trust A. Point of Trust A receives the confirming response and forwards it to the requesting communication party, client A.1.

Step 2: Setup of secure communication

The communication steps to setup secure communication are delineated subsequently.

Client A.1 and Point of Trust A are ready to communicate and generate Stream 1, a shared stream generated by using quantum mechanical techniques, over a quantum channel. Analog, Stream 2 is generated between Point of Trust A and client A.2 which is ready to communicate too.

Both communication parties have generated streams they share with Point of Trust A. Point of Trust A encrypts Stream 1 using Stream 2 to generate Stream 3. Stream 3 is then transmitted to client A.2.

Client A.2 receives Stream 3 and uses Stream 2, which is known to him, to reproduce Stream 1. Stream 1 is subsequently established as the session key for the communication parties, client A.1 and client A.2.

10.3.2 Speed-Oriented Setup of Communication

Contrary to the resource-oriented setup as described in 10.3.1, the response of the target, when requested to communicate, is not directly transmitted to the initiator. Instead the target responds by taking the initiative in setting up secure communication. In this way fewer steps are necessary, which speeds up the setup of communication.

10.3.2.1 Point of Trust Architecture with Several Trust Zones

In this system, communication takes place between clients from different Trust Zones. The respective Points of Trust must be trustworthy, i.e., they must have exchanged a key. For information on different network models see 10.3.1.1.

Step 1: Initiation of communication

At the beginning of the communication the situation is given by the following conditions. Client A.1 requests client B.2 to communicate and client A.1 and client B.2 are members of different Trust Zones.

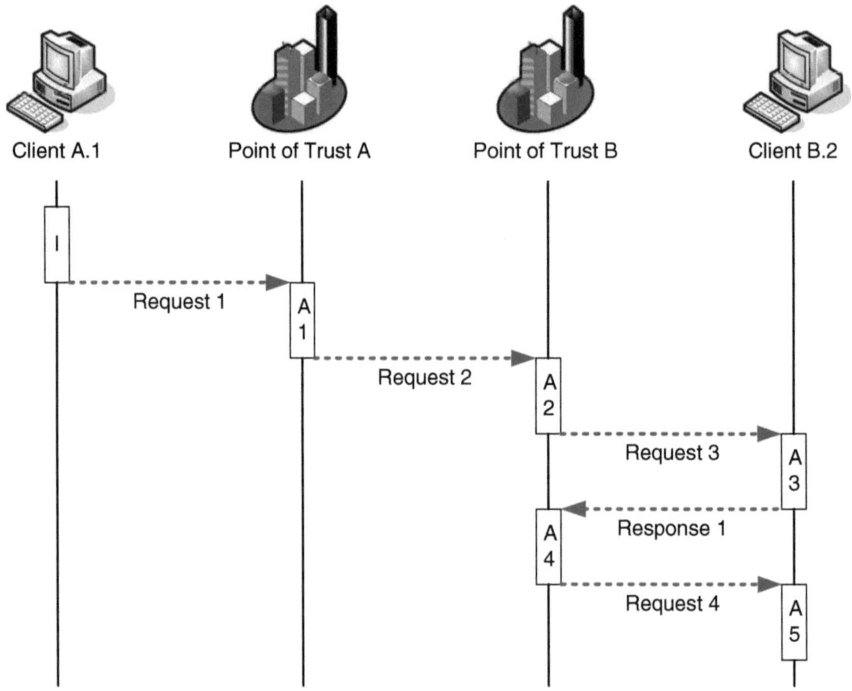

Fig. 10.7 Initiation of communication

The initialization of communication is shown in Fig. 10.7. The communication steps are delineated subsequently.

To initialize the communication client A.1 sends a request and transmits his data (authenticity, authorization, accounting information, etc.) to Point of Trust A (*I*) in order to be identified as part of Trust Zone A. Point of Trust A checks the data of client A.1. After positive validation, the request for communication party client B.2 is transmitted to Point of Trust B (*A1*) which is responsible for the target, client B.2.

Point of Trust B receives the request for communication between the communication parties client A.1 and client B.2 and requests the identification of client B.2 to communicate (*A2*). Client B.2 receives the request and sends his data (authenticity, authorization, accounting information, etc.) to Point of Trust B as response (*A3*) in order to be identified as part of Trust Zone B. Point of Trust B receives the response and checks the data of client B.2. After positive validation, the information relevant to the communication is sent to client B.2 (*A4*), introducing the requesting communication party, client A.1.

Client B.2 confirms the request for communication from client A.1 and takes initiative (*A5*).

Step 2: Setup of secure communication

The setup of secure communication is shown in Fig. 10.8. The communication steps are delineated subsequently.

Client A.1 and Point of Trust A are ready to communicate and generate Stream 1, a shared stream generated by using quantum mechanical techniques, over a quantum channel (*I1* and *I2*). Analog, client B.2, and Point of Trust B are ready to communicate and generate Stream 2 (*I3* and *I4*).

Both communication parties have generated streams they share with their respective Points of Trust. Point of Trust B encrypts Stream 2 using a key shared with Point of Trust A generating Stream 2 encrypted and transmits it to Point of Trust A (*A1*). Point of Trust A receives Stream 2 encrypted and uses the key shared with Point of Trust B to decrypt it. The reproduced Stream 2 is then encrypted by Stream 1 generating Stream 3 which is transmitted to client A.1 (*A2*).

Client A.1 receives Stream 3 and uses Stream 1, which is known to him, to reproduce Stream 2. Stream 2 is subsequently established as the session key for the communication parties, client A.1 and client B.2 (*A3*).

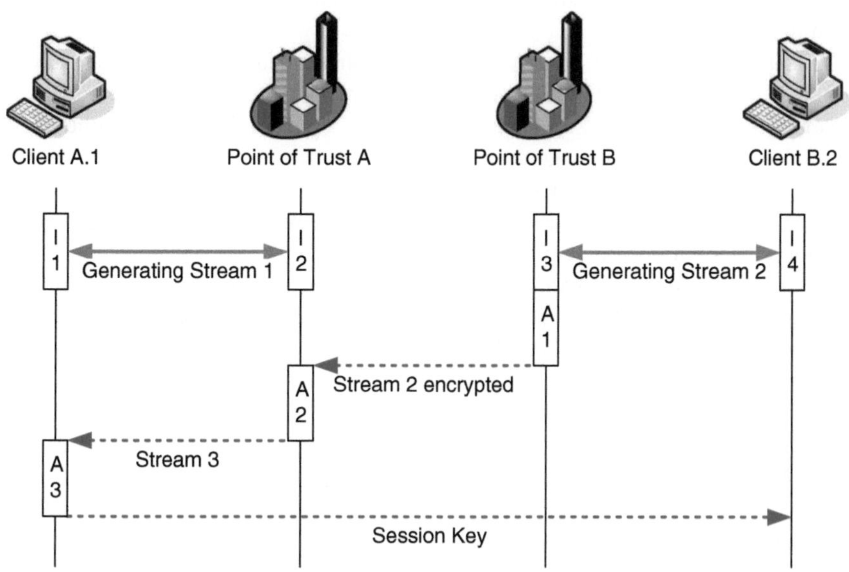

Fig. 10.8 Setup of secure communication

10.3.2.2 Point of Trust Architecture with One Trust Zone

Step 1: Initiation of communication

At the beginning of the communication the situation is given by the following conditions. Client A.1 requests client A.2 to communicate and client A.1 and client A.2 are members of the same Trust Zone.

The communication steps for the initialization of communication are delineated subsequently.

To initialize the communication client A.1 sends a request and transmits his data (authenticity, authorization, accounting information, etc.) to Point of Trust A in order to be identified as part of Trust Zone A. Point of Trust A checks the data of client A.1. After positive validation, Point of Trust A requests the identification of client A.2 to communicate.

Client A.2 receives the request and sends his data (authenticity, authorization, accounting information, etc.) to Point of Trust A as response in order to be identified as part of Trust Zone A. Point of Trust A receives the response and checks the data of client A.2. After positive validation, the information relevant to the communication is sent to client A.2, introducing the requesting communication party, client A.1.

Client A.2 confirms the request for communication from client A.1 and takes initiative.

Step 2: Setup of secure communication

The communication steps to setup secure communication are delineated subsequently.

Client A.1 and Point of Trust A are ready to communicate and generate Stream 1, a shared stream generated by using quantum mechanical techniques, over a quantum channel. Analog, Stream 2 is generated between Point of Trust A and client A.2 which is ready to communicate too.

Both communication parties have generated streams they share with Points of Trust A. Point of Trust A encrypts Stream 2 using Stream 1 to generate Stream 3. Stream 3 is then transmitted to client A.1.

Client A.1 receives Stream 3 and uses Stream 1, which is known to him, to reproduce Stream 2. Stream 2 is subsequently established as the session key for the communication parties, client A.1 and client A.2.

10.4 Exemplified Communications

The following section presents examples of communication models. It shows the difference between the setups for communications between several Trust Zones and within one Trust Zone and the setup for the generation of the bit stream itself.

10.4.1 Communication Between Several Trust Zones

In this system two Points of Trust are involved which are either connected directly or over an arbitrary network structure.

10.4.1.1 Initiation of Communication

At the beginning of the communication, it must be determined whether it is in the interest of the network operator and the two communication parties to set up communication. The request for communication is set up by preparing the required data in a packet.
The content of such a packet could be exemplified by the following:

- *Initiator*: The identification of the initiator, in this example the identification of client A.1, by a unique network address, which can be compared with other network protocols like IP.
- *Target*: Identification of the party who is requested to communicate, in this case the identification of client B.2.
- *Time* when communication is initiated: Determines the desired time of communication. Additionally, a token for an immediate setup of communication could be defined.
- *Conditions for the setup of communication*: Permits the definition of tokens regarding different communication factors. Examples of such tokens are payment (by initiator, target, other accounts, e.g., of a company), responsibility for the transmitted data, priority of communication, level of secrecy.

This request is transmitted over a public channel to the responsible Point of Trust, in this example Point of Trust A. In the first step the data of the initiator is examined. Such a check could include information as to, e.g., the authenticity, the authorization of the initiator to communicate (with the target, the level of secrecy, the point in time, etc.), and accounting information. If the validation fails, the communication is terminated and a response is generated accordingly.

If the identification is valid, then the information relating to the target is checked. The information check could include factors like availability of the target, level of secrecy of the communication. If the check yields a negative result, the communication is terminated and an appropriate message is transmitted to the initiator.

If the check yields a positive result, the request is forwarded to the Point of Trust, which is responsible for the target. This Point of Trust can be contacted in different ways. Examples are a hierarchical organization of the different Points of Trust or a direct connection of all Points of Trust.

The target's Point of Trust receives and processes the request. At this point, the data of the target is checked. Now the accounting information of the target could be checked, as this information may only be known to the respective client. Information regarding the level of secrecy or the ring of authorized communication parties could only be known to the responsible Point of Trust and might also be checked at this stage. If the check yields a negative result, the communication is terminated and an appropriate message is sent.

Additionally, the responsible Point of Trust could request the authentication of the target. At the same time, the target would be informed about the request for communication. Now all the information of the request can be transmitted. Alternatively, the data related to the request is forwarded after the positive identification, as depicted in Fig. 10.9. In this example the data required for communication is transmitted only after a valid identification of the target. This guarantees a higher level of security.

If the identification of the target fails, then the setup of communication is terminated and a response is sent accordingly. Otherwise the request is forwarded to the target. The target then decides whether to accept or reject the setup of communication. The response of the target is then routed to the initiator and to all Points of Trust involved. Depending on the form of communication the participants had previously agreed on (level of secrecy, accounting information, etc.) and the determined time, the required initializations can now be prepared.

In the case of a negative response, e.g., caused by the termination of communication, the initiator can start a new setup of communication. The Point of Trust may employ mechanisms to make use of operation parameters like, e.g., maximal number of attempts to set up communication over a certain period of time, costs per communication attempt.

Note: A higher level of security could be obtained by encrypting all the communication between the respective parties. Symmetric ciphers like block ciphers, stream ciphers, or asymmetric systems based on a central PKI would be suitable for that purpose.

10.4.1.2 Setup of Secure Communication

Once the communication is initiated the next step is to generate the required streams and to set up a secure communication as seen in Figs. 10.9 and 10.10. The initiator of the communication generates a stream with specified quality criteria and a respective key length with its Point of Trust. A key length of 256 bits is agreed upon to work subsequently with an encryption of, e.g., AES (see [2]).

As a next step the generated status is checked. A negative status leads to the termination of the communication. If the generation status is positive, the generated stream is encrypted by the initiator's Point of Trust. Symmetric ciphers like AES or asymmetric systems like PKIs can be used to encrypt the stream.

After receiving the encrypted Stream 1, the target's Point of Trust initiates the generation of a stream (Stream 2) with the target. Again different quality criteria can be agreed on. Depending on the encryption algorithm, that is subsequently used, the Stream 1 and Stream 2 have to be of equal length. Then the status is checked. In case of a negative status the communication is terminated.

If the status proves positive, then the encrypted Stream 1 is decrypted. As a next step Stream 1 is encrypted by Stream 2 to produce Stream 3. A One-Time-Pad cipher

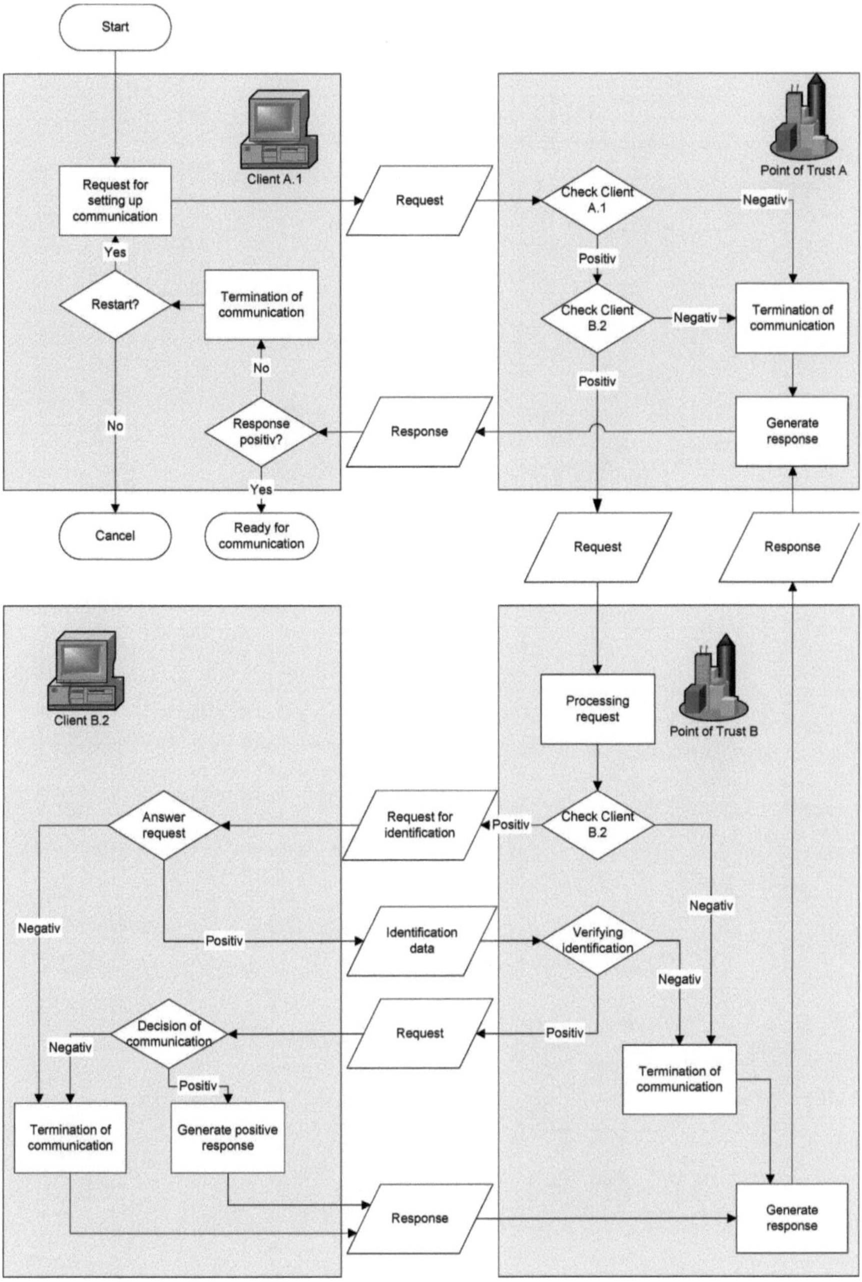

Fig. 10.9 Request for communication setup

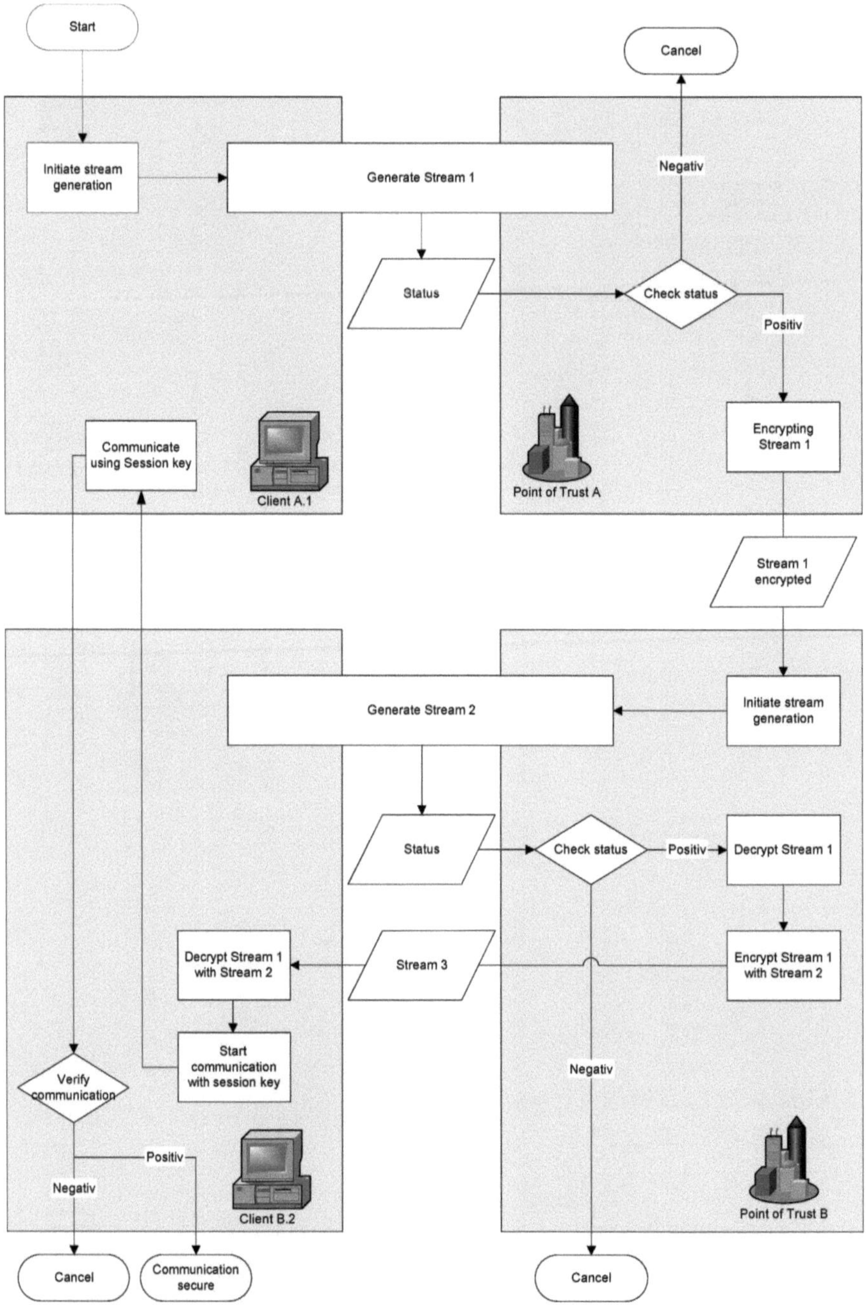

Fig. 10.10 Setup of secure communication

using the two streams is an example of a possible encryption. In this case, it is necessary that the streams are of equal length and fulfil predefined quality criteria.

Stream 3, which has been generated in this manner, is then transmitted to the target. The target decrypts Stream 3 using Stream 2. The result of the decryption (Stream 1) is subsequently referred to as the Session key.

The target now sets up communication with the initiator. Algorithms like AES or IDEA could be employed. Next, the setup of communication is checked. If the communication check yields a negative result, then the communication is terminated. If the result is positive, the communication is secure in the framework of the method.

Note: After the communication has been terminated, an appropriate message is sent to all of the participating communication parties. It is up to the system operator to define further steps in the event of the termination of a communication and to, e.g., repeat failed attempts. Whether a repeated attempt is permissible could depend on the requirements defined in the first step of the communication setup or on the prearranged time.

10.4.2 Communication in One Trust Zone

10.4.2.1 Initiation of Communication

At the beginning of the communication, it must be determined whether it is in the interest of the network operator and the two communication parties to set up communication. The request for communication is set up by preparing the required data in a packet.

The content of such a packet could be exemplified by the following:

- *Initiator:* The identification of the initiator, in this example the identification of client A.1, by a unique network address, which can be compared with other network protocols like IP.
- *Target:* Identification of the party who is requested to communicate, in this case the identification of client A.2.
- *Time when communication is initiated:* Determines the desired time of communication. Additionally a token for an immediate setup of communication could be defined.
- *Conditions for the setup of communication:* Permits the definition of tokens regarding different communication factors. Examples of such tokens are payment (by initiator, target, other accounts, e.g., of a company), responsibility for the transmitted data, priority of communication, level of secrecy.

This request is transmitted over a public channel to the responsible Point of Trust, in this example Point of Trust A. In the first step the data of the initiator is examined. Such a check could include information as to, e.g., the authentication of the initiator, authorization of the initiator to communicate (with target, the level of secrecy, the point in time, etc.), and accounting information. If the validation fails, the communication is terminated and an appropriate response is generated.

If the check yields a positive result, the information related to the target is examined. The check could include factors like, e.g., availability of the target, authorization to communicate with the initiator, accounting information of the target, level of secrecy. If the check yields a negative result, the communication is terminated and an appropriate message is forwarded to the initiator.

Additionally, the Point of Trust could request the identification of the target. At the same time, the target would be informed about the request for communication. Now all the information of the request can be transmitted. Alternatively, the data related to the request is forwarded after the positive identification, as depicted in Fig. 10.11. In this example the data required for communication is transmitted only after a valid identification of the target. This guarantees a higher level of security. If the identification fails, then the setup of communication is terminated and a response is sent accordingly.

Otherwise the request is forwarded to the target. The target then decides whether to accept or reject the setup of communication. The response of the target is then routed to the initiator and to the Point of Trust. Depending on the form of communication the participants had previously agreed on (level of secrecy, accounting information, etc.) and the determined time, the required initializations can now be prepared.

In the case of a negative response, e.g., caused by the termination of communication, the initiator can start a new setup of communication. The Point of Trust may employ mechanisms to make use of operation parameters like, e.g., maximal number of attempts to set up communication over a certain period of time, costs per communication attempt.

Note: A higher level of security could be obtained by encrypting all the communication between the respective parties. Symmetric ciphers like block ciphers (AES, IDEA, etc.), stream ciphers (one-time-pad, SEAL 2.0, etc.), or asymmetric systems based on a central PKI would be suitable for that purpose.

10.4.2.2 Setup of Secure Communication

Once the communication is initiated the next step is to generate the required streams and to set up a secure communication as shown in Figs. 10.11 and 10.12. The initiator of the communication generates a stream with specified quality criteria and a respective key length with his Point of Trust. For example, a key length of 256 bits is needed to work subsequently with AES (see [2]).

As a next step the status of the generation is checked. A negative status leads to the termination of the communication. If the status of the generation is positive, then the Point of Trust initiates the generation of a stream (Stream 2) with the target. Again, certain quality criteria (like [1]) can be agreed on. Depending on the encryption algorithm that is subsequently used, Stream 1 and Stream 2 have to be of equal length. Then the status is checked. In case of a negative status the communication is terminated. If the status proves positive, Stream 1 is encrypted by Stream 2 to

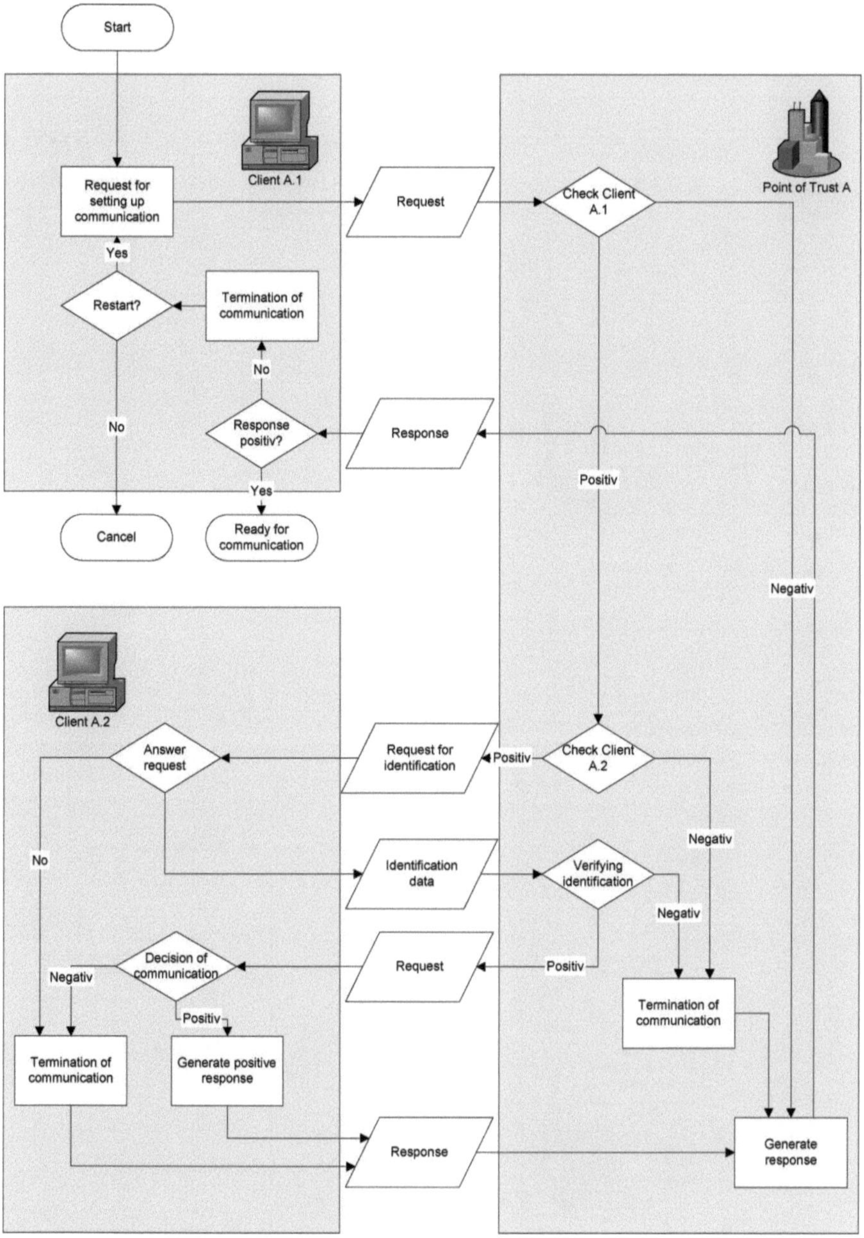

Fig. 10.11 Request for communication setup

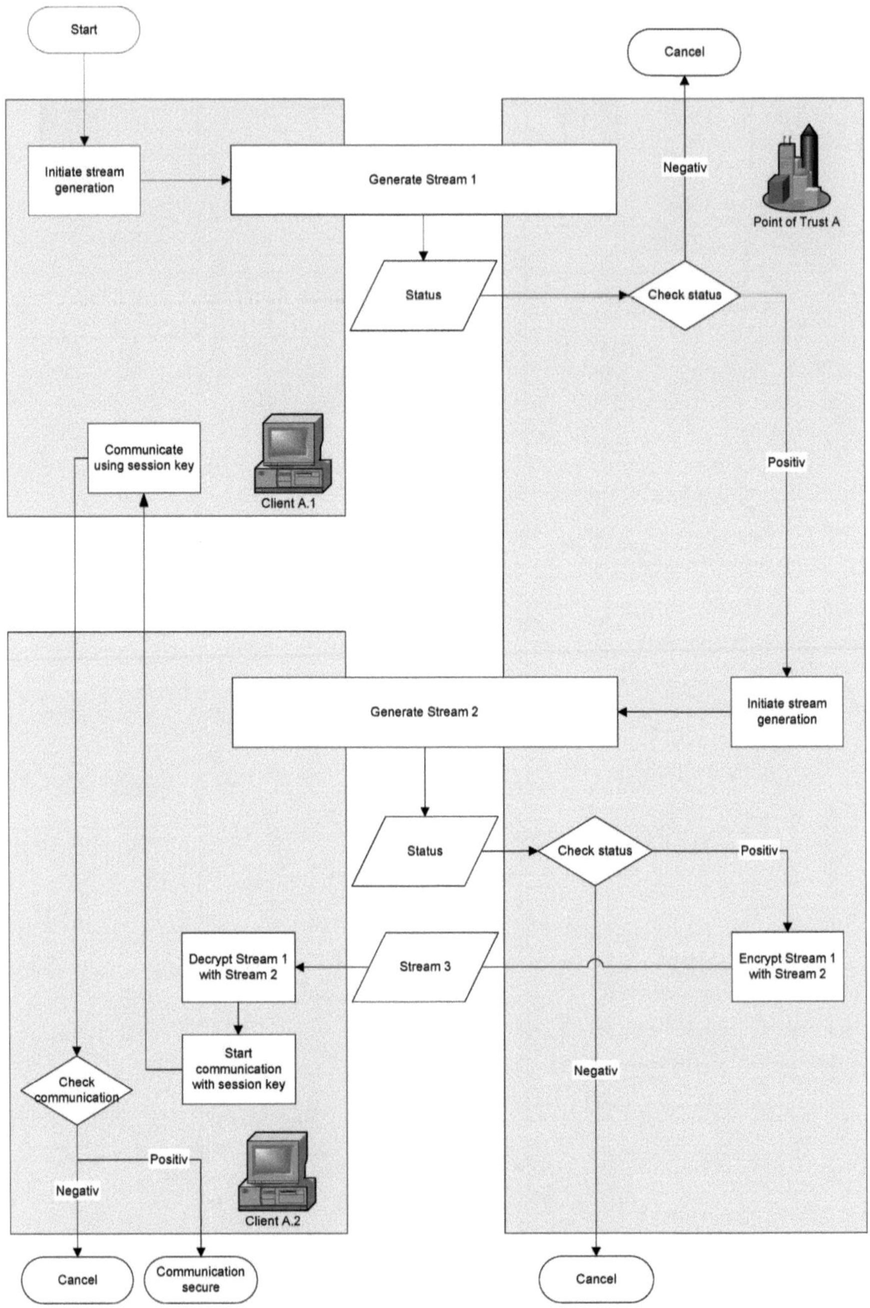

Fig. 10.12 Setup of secure communication

produce Stream 3. A One-Time-Pad cipher using the two streams is an example of a possible encryption. In this case, it is necessary that the streams are of equal length and fulfil predefined quality criteria.

Stream 3, which has been generated in this manner, is then transmitted to the target. The target decrypts Stream 3 by using Stream 2. The result of the decryption (Stream 1) is subsequently referred to as the Session key.

The target now sets up communication with the initiator. Algorithms like AES or IDEA could be employed. Next, the setup of communication is checked. If the communication check yields a negative result, then the communication is terminated. If the result is positive, the communication is secure in the framework of the method.

Note: After the communication has been terminated, an appropriate message is sent to all of the participating communication parties. It is up to the system operator to define further steps in the event of the termination of a communication and to, e.g., repeat failed attempts. Whether a repeated attempt is permissible could depend on the requirements defined in the first step of the communication setup or on the prearranged time.

10.4.3 Generation of a Stream

At the beginning of the generation process, the initiator prepares a detailed request as seen in Fig. 10.13. This request could include different tokens like the length of the stream that is to be generated, the quality criteria of the stream, e.g., FIPS 140-2 [1], criteria according to Golomb [9], linear complexity [9].

The request is forwarded to both communication parties and serves to initiate respective steps such as the preparation of the system (adjustments, self test, etc.), logging.

The participating communication parties generate a stream using a respective protocol, e.g., BB84 [3].

Once a stream with the specified length is generated, it is checked by one of the communication parties involved. In this check, the quality criteria specified in the request can be checked. If the test leads to the rejection of the generated stream, an appropriate error message is generated and sent to the second communication party.

If the stream is approved, a positive status message is generated. This message confirms that both communication parties now share a key, which has been generated by both parties.

It could be up to the initiator to make another attempt at generating a key if one try results in an error message. In this case, the system operator could interfere by, e.g., limiting the number of attempts over a certain period of time, charging each attempt separately.

If the attempt is not repeated, a negative status message is generated. It confirms that no shared key could be generated.

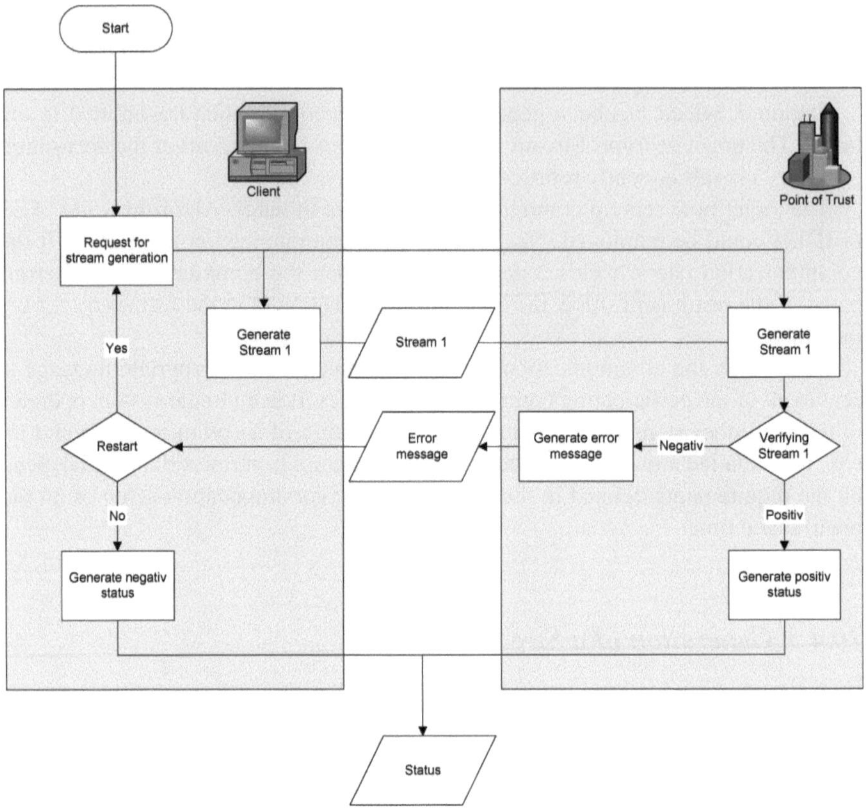

Fig. 10.13 Generation of a stream

10.5 A Medical Information System Based on the Ring of Trust

10.5.1 Field of Research

10.5.1.1 Availability on Demand

Besides the distribution of high-quality keys in an adequate amount also the storage of these keys is very important. In large data-processing centers it is possible to securely store such sensitive data since the required environment (access control, backup structure, hardware architecture) is present. But most of the commercial and private users do not have this kind of environment at their disposal. Thus we can state the requirement: keys should be generated on demand and not on supply to overcome the problems of their secure storage.

10.5.1.2 Scalability

Today we have different requirements about the security level of transmitted data. The data which gets transmitted during an online surgery over the Internet needs a much higher security level than most other data. Therefore it is necessary to draw a distinction between certain security levels for data to optimize speed and reduce the amount of needed resources.

Within a medical information system there exists data which could be encrypted with a classical algorithm for an adequate security level. There also exists data that is subject to legal requirements and hence needs to be encrypted with a certain algorithm. Further kinds of data need to be encrypted with predefined security algorithms to allow the use within existing environments or already used protocols like VPN. But beside this, most companies today have to maintain huge databases and transmit huge amounts of highly critical data, which needs high-level security.

10.5.1.3 The Problems of Distance and Speed

In our opinion it is another basic aspect that keys of adequate quality need to be available over long distances and with the according quality. QKD provides high-quality keys but today it is not possible to generate keys over very long distances. Within QKD systems increasing distance between Alice and Bob leads to decreasing transmission speed. So the deployment of classical QKD-based networks is limited.

10.5.2 Requirements

Medical Information Systems (MIS) [6, 8, 11] consisting of data-processing units like Radiology Information System (RIS), Lab Information System (LIS) [8], Digital Imaging and Communications in Medicine (DICOM), Health Level 7 (HL7) [10, 5], Picture Archiving and Communicating System (PACS) [6], ORBIS (MIS powered by GWI AG), International Statistical Classification of Diseases and Related Health Problems (ICD-10) [8] database or any other kind of digital information system are essential components of modern medical data processing units.

It is commonly known that patient-related data is highly sensitive. This kind of data needs to be stored for a longtime and its privacy has to be guaranteed for the whole time of storage. Only to a selected group of people like doctors or the nursing staff the access to the patient-related data should be granted. To prevent an intervention of an adversary it is essential to encrypt patient-related data during the transmission. Here it does not matter whether this information contains recipes, radiology reports, or any other kind of medical data.

Telemedicine is part of telematics-related surgeries [7] where a surgeon operates on a patient over a long distance – maybe even in another country. The surgeon's physiological transaction data is transmitted, for example, from Vienna to the operating theatre in Klagenfurt over a public channel. This highly sensitive data (e.g., heart surgeries over the Internet via triggering "da Vinci" surgery robot) needs to be protected against any adversary. Other important aspects besides high level of security are transmission speed and independence of the amount of transmitted data.

Manipulation of data is a severe threat and especially in medicine it could result in devastating consequences. Thus it is necessary to implement modern cryptographic methods into existing infrastructures. Additionally, techniques are needed which are easy to implement and also provide a high level of security and fulfil the requirements for transmission speed and long-range communication.

10.5.2.1 Results of the Survey

The survey [4] shows that the participants (medical technicians, commercial service providers, exc.) are very well informed about the state-of-the-art Medical Information Systems (cf. Fig. 10.14).

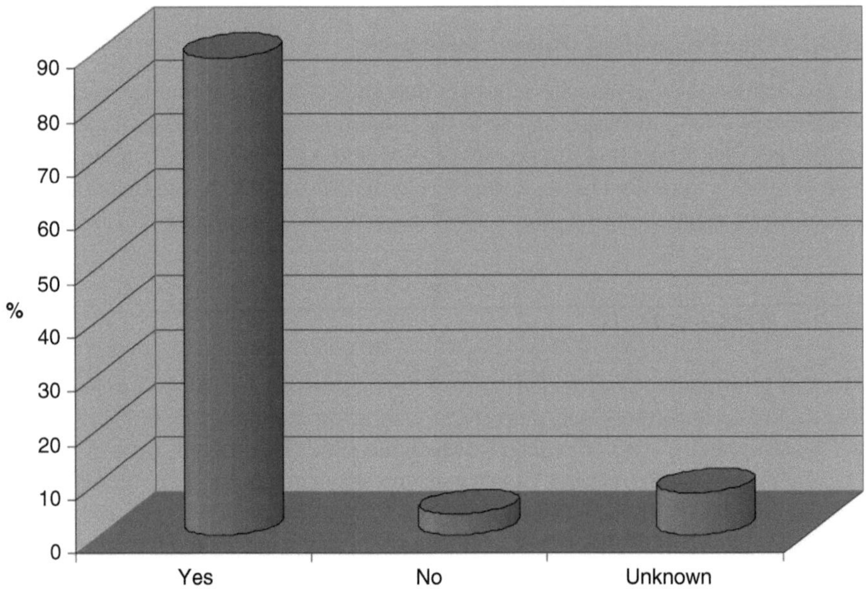

Fig. 10.14 Awareness about state-of-the-art Medical Information Systems

Regarding the privacy of transmitted data about 40% agreed that more than 75% of the communication has to be encrypted. Almost two-thirds say that more than 50% of the transmitted and stored data is sensitive (cf. Fig. 10.15).

As a result of the survey about 90% of the participants agreed that in medical data processing the need for high-quality security solutions is given. About 60% of the participants stress that current architectures could not meet the requirements stated in Sects. 10.5.1.1, 10.5.1.2, 10.5.1.3. They stated that modern solutions are needed to protect the patient's privacy to reach a high level of security (cf. Fig. 10.16).

QKD is a solution to these problems. It provides high-quality keys and fulfils the requirements from Sects. 10.5.1.1 and 10.5.1.2. A drawback of quantum key distribution is the range in which it is operable since it is rather low (about 20 km). But certain models like the Ring of Trust are able to solve also these problems.

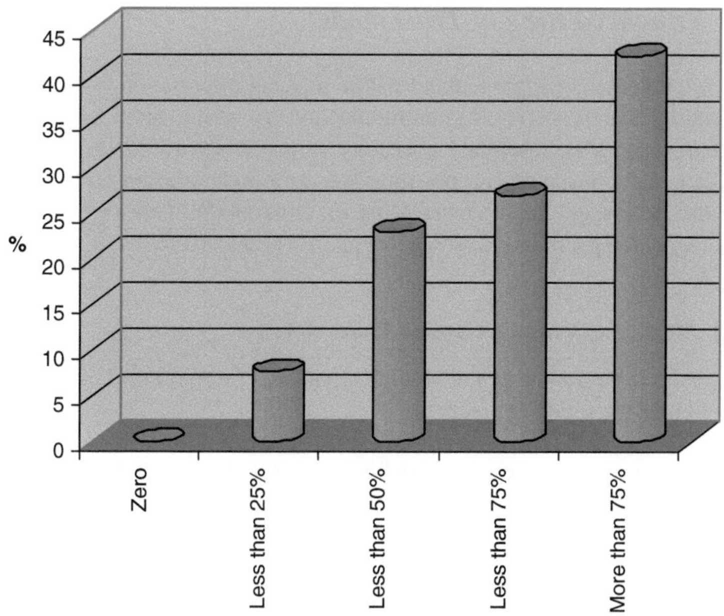

Fig. 10.15 The amount of sensitive data in Medical Information Systems

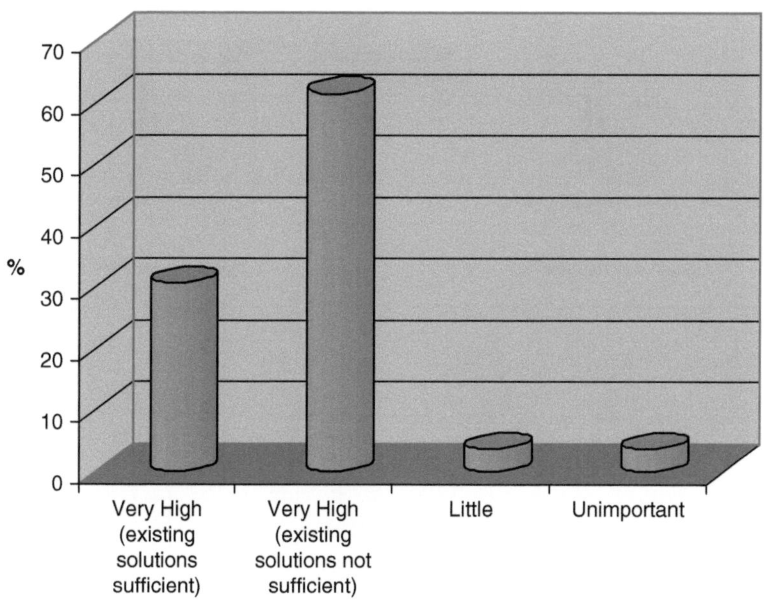

Fig. 10.16 The need for high-level security in Medical Information Systems

10.5.3 Enhanced Ring of Trust model

We can see that there is high demand within medical information systems, not only on generating shared secrets for communication but also to store information in a high secure way. It is absolutely necessary to get access to this highly protected information from any place at any time but only under certain circumstance. To enable this we present an enhanced Ring of Trust model with a special client by introducing a storage client as shown in Figs. 10.17 and 10.18.

10.5.3.1 Storage Client at a Foreign Point of Trust

To access the information on the storage client of a foreign Point of Trust a client A.1 has to take the following steps:

1. Client A.1 generates a key K_A with his Point of Trust. Client A.1 is a member of Trust Zone A and therefore assigned to Point of Trust A.
2. The key K_A is transmitted on a secure channel to the Point of Trust where the storage client is located, the Point of Trust B.
3. Point of Trust B generates key K_B with the storage client, encodes K_A with K_B to K_C (for example, by using the XOR operation) and sends this key either using the secure environment of Point of Trust B if the storage client is located within this secure environment or using a public channel.

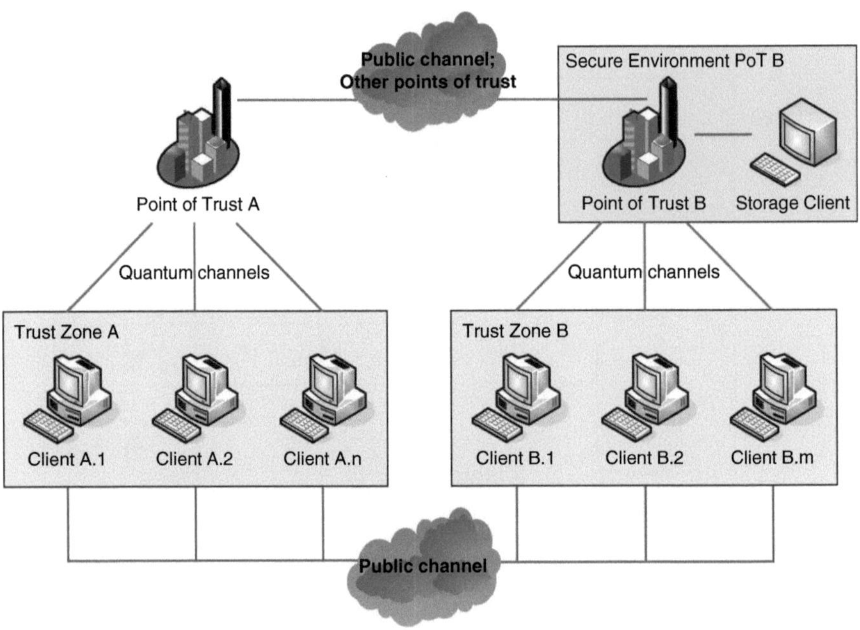

Fig. 10.17 Enhanced ring of trust model with storage client at a foreign point of trust

Fig. 10.18 Enhanced ring of trust model with storage client at the home point of trust

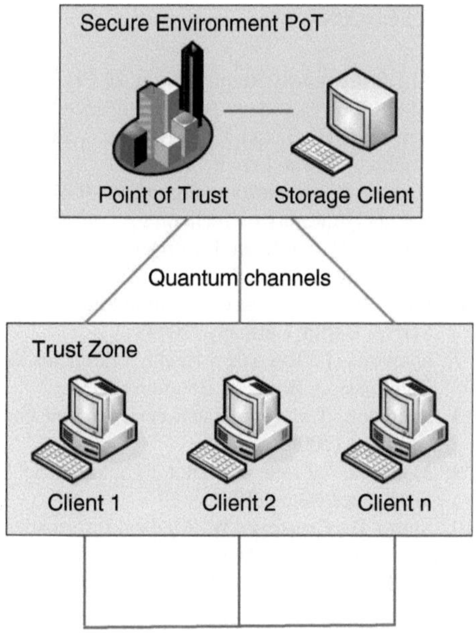

4. As the storage client knows K_B he is able to obtain K_A from K_C by calculating $K_A = K_C \oplus K_B$.

With K_A the storage client is now able to establish a secure connection to client A.1. If the storage client is located within the secure environment of the Point of Trust, it would be possible to transmit K_A directly to the storage client without the generation of K_B.

10.5.3.2 Storage Client at the Home Point of Trust

To get access to the information on the storage client at the same trust center, a client has to take the following steps:

1. Client A.1 generates a key K_A with his Point of Trust. Client A.1 is a member of Trust Zone A and therefore assigned to Point of Trust A.
2. If the storage client is located within the secure environment of Point of Trust A, key K_A could be transmitted directly to the storage client.
3. If the storage client is not located within the secure environment but within the Trust Zone of Point of Trust A a key K_B can be generated by the Point of Trust and the storage client.
4. The Point of Trust encodes a new key K_C by calculating $K_C = K_A \oplus K_B$ and sends this key to the storage client by using a public channel.
5. As the storage client knows K_B he is able to calculate K_A from K_C.

Using K_A the storage client is now able to establish a secure connection to client A.1.

References

1. F.I.P.S.P. Security Requirements for Cryptographic Modules. NIST, 140-2
2. F.I.P.S.P, 197: Announcing the Advanced Encryption Standard (AES). NIST
3. Bennet, C.H., Besset, F., Brassard, G., Salvail, L., Smolin, J.: Experimental Quantum Cryptography. J. Cryptology **5**, 3(1992)
4. Dissauer, G.: Security requirements of modern medical information systems. Master's thesis, Vienna University of Technology, Austria (2007)
5. Dolin, R.H., Aschuler, L., Boyer, S., Beebe, C., Behlen, F.M., Biron, P.V., Shabo, A.: Medical Informatics Europe. J. Am. Med. Inform Assoc. (2006)
6. Haas, P.: Medizinische Informationssysteme und elektronische Krankenakten. Springer Verlag, Berlin, Germany (2004)
7. Iakovidis, I.: Towards a Health Telematics Infrastructure in the European Union. European Commission, Brussels, Belgium (2000)
8. Lehmann, T.M.: Handbuch der Medizinischen Informatik. Carl Hanser Verlag, Munchen, Germany (2005)
9. Menezes, A.J., van Oorschot, P.C., Vanstone, S.A.: Handbook of Applied Cryptography. CRC Press Boca Raton (1996)
10. Smith, B., Ceustersc, W.: Medical Informatics Europe. Maastricht, Netherlands (2006)
11. van Bemmel, J., Musen, M.A.: Handbook pf Medical Informatics. Springer Verlag, Heidelberg, Germany (1997)

Index

Kollmitzer, C., Pivk, M. (eds.): *Index*. Lect. Notes Phys. **797**, 211–214 (2010)
DOI 10.1007/978-3-642-04831-9 © Springer-Verlag Berlin Heidelberg 2010

MIX
Papier aus verantwortungsvollen Quellen
Paper from responsible sources
FSC® C105338

If you have any concerns about our products,
you can contact us on
ProductSafety@springernature.com

In case Publisher is established outside the EU,
the EU authorized representative is:
Springer Nature Customer Service Center GmbH
Europaplatz 3, 69115 Heidelberg, Germany

Printed by Libri Plureos GmbH
in Hamburg, Germany